Use of Progestogens
in
Clinical Practice of Obstetrics and Gynecology

Use of Progestogens in Clinical Practice of Obstetrics and Gynecology

Anita MV MD MNAMS MICOG FICMCH
Consultant
Obstetrics and Gynecology

Sandhya Jain MD DNB MNAMS FICOG
Associate Professor
Department of Obstetrics and Gynecology
University College of Medical Sciences and
Guru Teg Bahadur Hospital
Delhi, India

Neerja Goel MD FAMS FICOG FIMSA FFGSI
Ex Director Professor
Department of Obstetrics and Gynecology
University College of Medical Sciences and
Guru Teg Bahadur Hospital
New Delhi, India

Foreword
Amita Suneja

JAYPEE *The Health Sciences Publisher*
New Delhi | London | Panama

Jaypee Brothers Medical Publishers (P) Ltd.

Headquarters
Jaypee Brothers Medical Publishers (P) Ltd.
4838/24, Ansari Road, Daryaganj
New Delhi 110 002, India
Phone: +91-11-43574357
Fax: +91-11-43574314
E-mail: jaypee@jaypeebrothers.com

Overseas Offices

J.P. Medical Ltd.
83, Victoria Street, London
SW1H 0HW (UK)
Phone: +44-20 3170 8910
Fax: +44(0)20 3008 6180
E-mail: info@jpmedpub.com

Jaypee-Highlights Medical Publishers Inc.
City of Knowledge, Bld. 235, 2nd Floor, Clayton
Panama City, Panama
Phone: +1 507-301-0496
Fax: +1 507-301-0499
E-mail: cservice@jphmedical.com

Jaypee Brothers Medical Publishers (P) Ltd.
17/1-B, Babar Road, Block-B, Shaymali
Mohammadpur, Dhaka-1207
Bangladesh
Mobile: +08801912003485
E-mail: jaypeedhaka@gmail.com

Jaypee Brothers Medical Publishers (P) Ltd.
Bhotahity, Kathmandu, Nepal
Phone: +977-9741283608
E-mail: kathmandu@jaypeebrothers.com

Website: www.jaypeebrothers.com
Website: www.jaypeedigital.com

© 2018, Jaypee Brothers Medical Publishers

The views and opinions expressed in this book are solely those of the original contributor(s)/author(s) and do not necessarily represent those of editor(s) of the book.

All rights reserved. No part of this publication may be reproduced, stored or transmitted in any form or by any means, electronic, mechanical, photocopying, recording or otherwise, without the prior permission in writing of the publishers.

All brand names and product names used in this book are trade names, service marks, trademarks or registered trademarks of their respective owners. The publisher is not associated with any product or vendor mentioned in this book.

Medical knowledge and practice change constantly. This book is designed to provide accurate, authoritative information about the subject matter in question. However, readers are advised to check the most current information available on procedures included and check information from the manufacturer of each product to be administered, to verify the recommended dose, formula, method and duration of administration, adverse effects and contraindications. It is the responsibility of the practitioner to take all appropriate safety precautions. Neither the publisher nor the author(s)/editor(s) assume any liability for any injury and/or damage to persons or property arising from or related to use of material in this book.

This book is sold on the understanding that the publisher is not engaged in providing professional medical services. If such advice or services are required, the services of a competent medical professional should be sought.

Every effort has been made where necessary to contact holders of copyright to obtain permission to reproduce copyright material. If any have been inadvertently overlooked, the publisher will be pleased to make the necessary arrangements at the first opportunity. The **CD/DVD-ROM** (if any) provided in the sealed envelope with this book is complimentary and free of cost. **Not meant for sale.**

Inquiries for bulk sales may be solicited at: jaypee@jaypeebrothers.com

Use of Progestogens in Clinical Practice of Obstetrics and Gynecology

First Edition: **2018**
ISBN: 978-93-5270-218-3

Dedication

We dedicate this book to all the teachers who impart knowledge and wisdom to their students and show them the path of righteousness, happiness and success.

Foreword

Medical students are always in need for comprehensive books in relevant topics to not only aid them in their examinations but also to enhance clinical skills.

This book *Use of Progestogens in Clinical Practice of Obstetrics and Gynecology* endeavors to help the students and practicing gynecologists in understanding each and every aspect of Progestogens, which are being prescribed so often. It is ardently hoped that this excellent publication will be profitably used by the readers in improving their clinical knowledge and hence patient care. Chapters in this book range from basic principles, nomenclature, pharmacodynamics, description of individual progestogen and their clinical utility. Chapter fifteen has been written with the idea to expertise in prescription writing.

The learned editors have done an admirable job and deserve high appreciation for their devoted efforts.

In my long professional career, I have never felt so proud to see the growth and potential of the editors and have the opportunity and honor to recognize it.

Amita Suneja
Director Professor and HOD
Department of Obstetrics and Gynecology
University College of Medical Sciences and
Guru Teg Bahadur Hospital, Delhi, India

Preface

The subject of Gynecology and Obstetrics includes a variety of hormones, most important being Estrogen and Progesterone which regulate the female physiology immensely. A complete understanding of these hormones is of paramount importance for correct prescription writing.

We have felt the need for a comprehensive book on Progesterone and Progestogens to aid the postgraduate students in their medical training and examinations. Also for practicing gynecologists, an update on type, route and regimes of Progestogens have been incorporated in the manual.

This book has sixteen elaborate chapters for complete understanding of the natural molecule 'Progesterone' and also the most recently launched synthetic 'Progestins or Progestogens' e.g. Gestodene, Dienogest, Nesterone, Drosperinone etc. There are ample illustrations for better understanding. The first of its kind, this book covers all the aspect of native molecule progesterone and its synthetic analog Progestogens. This book aims to clarify the doubts about its pharmacokinetics, clinical uses, dosage and contraindications; This will rationalize the treatment protocols.

We hope this book will help the readers to be able to use Progestins more judiciously in their clinical practice and a feedback to this book would be welcomed.

Anita MV
Sandhya Jain
Neerja Goel

Acknowledgments

The editors are indebted to Dr Sunil Kumar—Medical Director, Guru Teg Bahadur Hospital and Dr VP Gupta—Principal, University College of Medical Sciences, Delhi for their encouragement and support.

We offer our sincere gratitude to Director Professor Amita Suneja, Head of Department of Obstetrics and Gynecology, Guru Teg Bahadur Hospital, Delhi for allowing and encouraging us to pursue our academic venture.

No work of excellence can be done in isolation. We would like to thank the faculty and residents of the Department of Obstetrics and Gynecology, Guru Teg Bahadur Hospital, Delhi for their unconditional support.

No words of praise would be adequate enough for the entire staff of M/s Jaypee Brothers Medical Publishers, New Delhi for their assistance and support.

In the end, we show our gratitude to our family members who gave us their full cooperation and supported us in this endeavor.

Anita MV
Sandhya Jain
Neerja Goel

Contents

1. Introduction and History 1
2. Structure, Biochemistry and Biosynthesis 5
3. Nomenclature and Classification 8
4. Routes of Administration and Bioavailability 11
5. Micronization 15
6. Pharmacokinetics and Metabolism 17
7. Physiological Actions of Progesterone 21
8. Pharmacodynamics and Selectivity of Progestins 24
9. Individual Progestogen Description 27
10. Clinical Usage Guidelines of Progestins 35
11. Role of Progestins in
    Combined Hormonal Contraception 62
12. Side Effects, Contraindications and Precautions 67
13. Patient Advice 76
14. Drug and Test Interactions 77
15. Prescription Writing 79
16. References 84

Index *89*

CHAPTER 1

Introduction and History

Progesterone is an endogenous steroid and sex hormone involved in the menstrual cycle, pregnancy and embryogenesis of humans and other species. Progesterone is a natural substance, which is produced by the corpus luteum for maintenance of pregnancy. It acts on the endometrium, decreases estrogen receptors and mitotic activity and develops the stromal component of the endometrium. This prepares the endometrium for future prostaglandin production, permitting complete uniform shedding at the time of progesterone withdrawal. In larger doses it decreases the effect of estrogen which includes epithelial stimulation of endometrium, breast and metabolic activities. The parent compound has limitations for clinical utility because of poor absorption, rapid hepatic inactivation and transient intramuscular action. Various progesterone like or progestational agents have been synthesized and called progestogens, progestins or gestagens. While the progestins are structural analogues of progesterone, they are not functional analogues, and differ from progesterone considerably in their properties.[1,2]

Progestogens, are a class of steroid hormones that bind to and activate the progesterone receptor (PR).[3] The progestogens are named for their function in maintaining pregnancy (i.e. *progestational*), although they are also present at other phases of the estrous and menstrual cycles.[4,5] They are one of three types of sex hormones, the others being estrogens like estradiol and androgens/anabolic steroids like testosterone. The progestogens are one of the five major classes of steroid hormones, in addition to the androgens, estrogens, glucocorticoids, and mineralocorticoids, as well as the neurosteroids. All endogenous progestogens are characterized by their basic 21-carbon skeleton, called a pregnane skeleton (C21). In similar manner, the estrogens possess an estrane skeleton (C18), and androgens, an androstane skeleton (C19).

The most important progestogen in the body is progesterone (P4).[6,7] Other endogenous progestogens include 16α-hydroxyprogesterone,

17α-hydroxyprogesterone, 20α-dihydroprogestero, 5α-dihydroprogesterone, 11-deoxycorticosterone, and 5α-dihydrodeoxycorticosterone.[8-13]

Major examples of progestins include the 17α-hydroxyprogesterone derivative medroxyprogesterone acetate and the 19-nortestosterone derivative norethisterone (norethindrone).

It is on the WHO model list of "Essential Medicines", the most important medications needed in a basic health system. Progestins are widely used for various indications from menarche to pregnancy and menopause. A wide variety of progestational compounds are existing in the market and different agents are used for different medical conditions and indications. This monogram is an endeavor to elucidate the rational choice and use of progestins in various gynecological disorders.

It was in 1923 that the hormonal action of progesterone was discovered.[14-16] It was then known by the name of luteosterone, as it was known to be produced by the corpus luteum of ovaries. Purification of progesterone was the work of several investigators between 1928 and 1934. WM Allen, an American gynecologist and GW Corner, his Anatomy professor isolated the hormone progesterone in 1929 while working in embryology laboratory at the University of Rochester. They are credited with the discovery and the first isolation of pure progesterone from the waxy material obtained by high-vacuum distillation from the corpus luteum extracts of animals. In 1930 they pioneered and used corpus luteum extracts to maintain pregnancy after early ablation of rabbit ovaries.

By 1932, crystalline material of high progestational activity had been isolated from the corpus luteum of animals. By 1934, the extraction of progesterone from urine was accomplished and the chemical structure of progesterone was determined by professor Adolf Butenandt at the Chemisches Institute of Technical University in Gdańsk, Poland.[14,15,17] Later that year it was chemically synthesized from stigmasterol and pregnanediol.[15,18] Professor Butenandt received the Nobel Prize for Chemistry in 1939 for the discovery and identification of progesterone, estrogen and androsterone. He initially rejected the award in accordance with government policy, but accepted it in 1949 after World War II.

In 1935, in the Second International Conference on the Standardization of Sex Hormones in London, England, organized under the chairmanship of Sir Henry Dale to discuss the terminology and standard tests for the "second" female sex hormone, the name progesterone (derived from the three words—progestational steroidal ketone) was accepted for common use in scientific literature.[14,19] Shortly after this, progesterone began being tested clinically in women.[15] In 1934, Schering introduced progesterone, as a pharmaceutical drug, under the brand name Proluton, along with estradiol (brand name Progynon) and testosterone (brand names Testoviron, Proviron).[20,21] The parent compound had its limitations for clinical utility viz. poor oral absorption, rapid hepatic inactivation and transient intramuscular action. With the aim to overcome these limitations, synthesis of progesterone-like or progestational agents was attempted and 17-hydroxy-progesterone caproate was the first progestin obtained. It required parenteral administration.

In 1943, Russel Marker, an American chemist, synthesized progesterone from diosgenin derived from Mexican yam roots which considerably brought down its cost. In 1950, synthesis of C-21 steroid medroxyprogesterone acetate and related compounds megestrol acetate and chlormadinone yielded highly potent and orally active agents. It was demonstrated in 1951 that removal of the 19-carbon from ethisterone to form norethindrone changed its hormonal effect from androgenic to that of a progestational agent which is orally active. These compounds were appropriately termed as 19-nortestosterones and they were not totally free from androgenic and anabolic properties. In 1952, chemists in USA and Mexico synthesized derivatives of 17-ethinyl-19-nortestosterone, which yielded powerful orally active progestogens. Substitutions of an 18-ethyl group are among the most widely used progestogens today.

The process of micronization of progesterone allowed its better and more effective absorption via other routes of administration. Oral micronized progesterone was marketed in 1980 in France under the brand name Utrogestan and this was followed by the introduction of oral micronized progesterone in the United States under the brand name Prometrium in 1998.[21-24] In the early 1990s, vaginal micronized progesterone (brand names Crinone, Utrogestan, Endometrin) was also marketed.[25,26]

Progesterone was approved by the United States Food and Drug Administration (USFDA) as vaginal gel on July 31, 1997, an oral capsule on May 14, 1998 in an injection form on April 25, 2001 and as a vaginal insert on June 21, 2007.[27-30]

Progesterone is marketed under a large number of different brand names throughout the world.[31] Progesterone, when taken orally, has very poor pharmacokinetics, including low bioavailability (only about

10–15% reaches the bloodstream) and a half-life of only about 5 minutes, unless it is micronized.[7,11,32] As such, it is sold in the form of oil-filled capsules containing micronized progesterone for oral use (Utrogestan, Prometrium, Microgest), termed oral micronized progesterone (OMP).[7,31]

Progesterone is also available in the forms of vaginal or rectal suppositories or pessaries (Cyclogest), transdermally-administered gels or creams (Crinone, Endometrin, Progestogel, Prochieve), or via intramuscular or subcutaneous injection of a vegetable oil solution (Progesterone, Strone).[7,31,33,34]

Transdermal products made with progesterone United States Pharmacopeia (i.e. "natural progesterone") do not require a prescription. Some of these products also contain "wild yam extract" derived from Dioscorea villosa, but there is no evidence that the human body can convert its active ingredient (diosgenin, the plant steroid that is chemically converted to produce progesterone industrially) into progesterone.[35-37]

CHAPTER 2

Structure, Biochemistry and Biosynthesis

The chemical name of progesterone is pregn-4-ene-3,20-dione. It belongs to a group of steroid hormones called the progestogens and is the major progestogen in the body.[11] Progesterone is also a crucial metabolic intermediate in the production of other endogenous steroids including the sex hormones and corticosteroids and plays an important role in brain function as a neurosteroid.[12]

Progesterone occurs as a white or creamy white, crystalline powder. It is odorless and is stable in air. Practically insoluble in water, it is soluble in alcohol, acetone, and dioxane and sparingly soluble in vegetable oils.

It has the following structural formula:

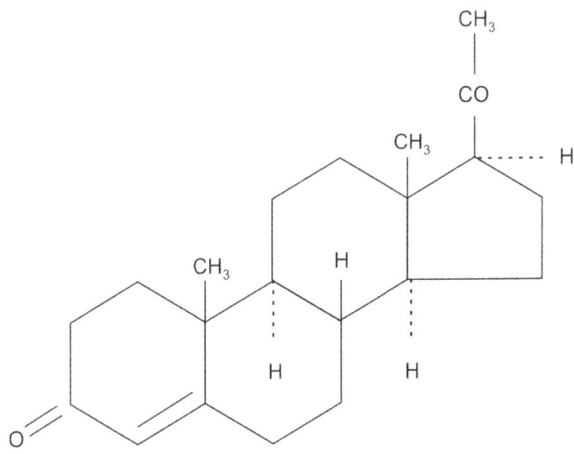

Pregn-4-ene-3,20-dione
Chemical data:
Formula: $C_{21}H_{30}O_2$
Molar mass: 314.46 g/mol

Systematic (IUPAC) name:
(8S,9S,10R,13S,14S,17S)-17-acetyl-10, 13-dimethyl-1,2,6,7,8,9,11,12, 14,15,16,17-dodecahydrocyclopenta[a]phenanthren-3-one.

Progesterone injection, is a sterile solution of Progesterone in a suitable vegetable oil available for intramuscular use. Each mL contains: Progesterone 50 mg, Benzyl alcohol 10% as preservative in Sesame oil q.s.

Progesterone is formed from steroid precursors in the ovaries, testes, adrenal cortex, and placenta. It is secreted mainly from the corpus luteum of the ovary during the latter half of the menstrual cycle. In the non-pregnant woman, peripheral conversion of steroids to progesterone is not seen. Pregnenolone and progesterone can also be synthesized by yeast.

In mammals, progesterone like all other steroid hormones, is synthesized from pregnenolone, which in turn is derived from cholesterol.[3,5,6]

Cholesterol undergoes double oxidation to produce 20,22-dihydroxycholesterol (Flowchart 2.1). This vicinal diol is then further oxidized with loss of the side chain starting at position C-22 to produce pregnenolone.[3-5] This reaction is catalyzed by cytochrome P450scc. The conversion of pregnenolone to progesterone takes place in two steps. First, the 3-hydroxyl group is oxidized to a keto group and second, the double bond is moved to C-4, from C-5 through a keto/enol tautomerization reaction.[1,38] This reaction is catalyzed by 3β-hydroxysteroid dehydrogenase/δ(5)-δ(4)isomerase. Progesterone in turn is the precursor of the mineralocorticoid aldosterone, and after conversion to 17α-hydroxyprogesterone (another natural progestogen) of cortisol and androstenedione. Androstenedione can be converted to testosterone, estrone and estradiol.

Flowchart 2.1: Steroidogenesis.

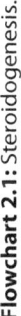

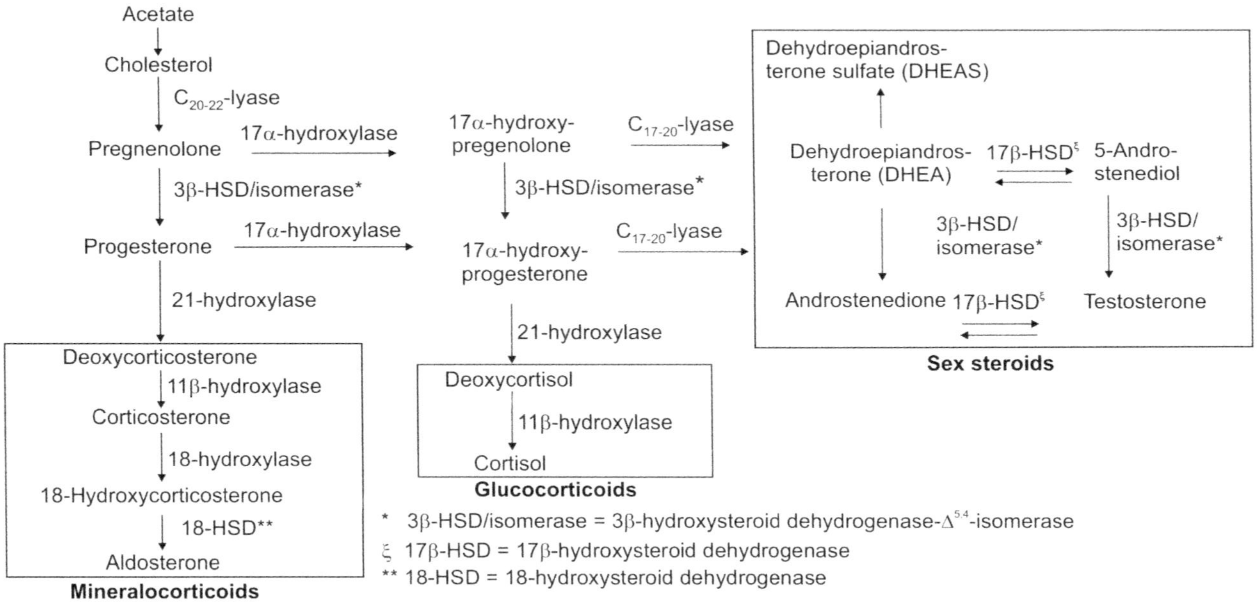

* 3β-HSD/isomerase = 3β-hydroxysteroid dehydrogenase-$\Delta^{5,4}$-isomerase

ξ 17β-HSD = 17β-hydroxysteroid dehydrogenase

** 18-HSD = 18-hydroxysteroid dehydrogenase

Source: Murray RK, Bender DA, Botham KM, Kennelly PJ, Rodwell VW, Weil P. Harper's Illustrated Biochemistry, 29e; 2012. Available at: http://accessmedicine.mhmedical.com/content.aspx?bookid=389&sectionid =40142497&jumpsectionID =40144300

CHAPTER 3

Nomenclature and Classification

CLASSIFICATION BY ORIGIN

Progestogens can be broadly classified into natural and synthetic forms (Flowchart 3.1).

Natural

The only natural progestin is progesterone. It is obtained from animal ovaries, Mexican yam roots and soybeans.

Synthetic Progestins

These can be further classified as followings:
1. *Progestogens structurally related to progesterone*: These compounds are obtained from the manipulation of progesterone molecules. Addition of hydroxyl group at C-17 followed by acetylation of this group gives rise to 17-hydroxyprogesterone acetate. Stereoisomerization of the progesterone molecule gives rise to dydrogesterone. Various oral and parenteral compounds are obtained by manipulation of C-6 carbon atom viz. medroxyprogesterone acetate, megestrol acetate, cyproterone acetate, chlormadinone acetate. All these compounds come under the class of Pregnanes.
2. *Progestogens structurally related to testosterone*: These include Estranes and Gonanes. These are derived from 19-nortestosterone. Addition of ethinyl group at C-17 of testosterone causes the steroid to lose androgenicity and to acquire progestational properties and oral activity. Removal of methyl group at C-10 further increases the oral progestational activity. The resulting estrane derivative is called norethindrone or norethisterone. Progestogens resembling norethindrone are converted to more potent lynestrenol, norethindrone acetate and ethynodiol diacetate.

3. *Substitution of ethyl group in place of methyl group*: This substitution of methyl group at C-13 gives rise to a more potent oral progestogen known as norgestrel, the active form of this being levonorgestrel. Manipulation of levonorgestrel gives rise to norgestimate and desogestrel. This group of compounds is called Gonanes.
4. *Newer progestins*: These are 19-norprogesterone derivatives and include Dienogest, Promegestone, Nestorone, Nomegestrol, Drospirenone and Trimegestone.

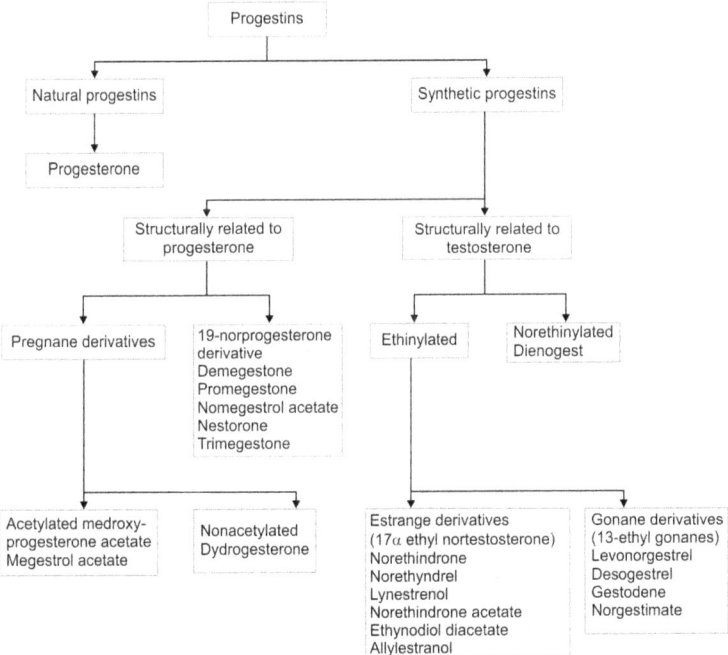

Flowchart 3.1: Classification of progestogens by origin

CLASSIFICATION BY CLINICAL UTILITY

1. *Pregnanes*: These are C-21 compounds structurally related to progesterone. Examples are: 17-hydroxyprogesterone acetate, 17-hydroxyprogesterone caproate, dydrogesterone, medroxyprogesterone acetate, cyproterone acetate and megestrol acetate.

2. *Estranes*: These are C-18 compounds and 19-nortestosterone derivatives. They are more potent and more androgenic. Examples are: norethindrone, norethynodrel, allylestrenol, lynestrenol.
3. *Gonanes*: These are C-19 compounds and ethinylated derivatives of estranes. They are the most potent and selective compounds. Examples are: levonorgestrel, norgestimate, gestodene, desogestrel.

CHAPTER 4

Routes of Administration and Bioavailability

PHYSIOLOGICAL BLOOD LEVELS

In women, progesterone is produced from the ovaries and the adrenals, with the adrenals contributing a very minor fraction. The blood progesterone levels are relatively low during the preovulatory phase of the menstrual cycle, rise after ovulation, and are elevated during the luteal phase, as shown in diagram below. Progesterone levels tend to be <2 ng/mL prior to ovulation, and >5 ng/mL after ovulation. The levels range from 3 to 15 ng/mL during the luteal phase. The blood production rate is less than 1 mg/day in the preovulatory phase, which increases postovulation to 20-30 mg/day. If pregnancy occurs, human chorionic gonadotropin is released maintaining the corpus luteum allowing it to maintain levels of progesterone. Between 7 and 9 weeks the placenta begins to produce progesterone in place of the corpus luteum, this process is named the luteal-placental shift.[39] After the luteal-placental shift progesterone levels start to rise further and may reach 100-200 ng/mL at term. Whether a decrease in progesterone levels is critical for the initiation of labor has been argued and may be species-specific. After delivery of the placenta and during lactation, progesterone levels are very low.

Progesterone levels are relatively low in children and postmenopausal women.[40] Adult males have levels similar to those in women during the follicular phase of the menstrual cycle. Blood progesterone levels can be 50 times above normal in congenital adrenal hyperplasia. Blood test results should always be interpreted using the reference ranges provided by the laboratory that performed the results. Example reference ranges are listed below.

For comparative purposes, mid-luteal serum levels of progesterone are above 5-9 ng/mL, plasma levels in the first 4 to 8 weeks of pregnancy are 25-75 ng/mL, and serum levels at term are typically around 200 ng/mL. Production of progesterone in the body in late pregnancy is approximately 250 mg per day, 90% of which reaches maternal circulation.[14,41,42]

ROUTES OF ADMINISTRATION

The different routes of administration for progesterone are orally, parenterally, intravaginally as a gel or insert and transdermal or subcutaneous.[43-47]

Oral Administration

- *Progesterone*: Following oral administration, it undergoes extensive first pass metabolism in gut wall and liver so that only ¼ of the dose reaches the circulation. The peak plasma levels are obtained in two hours that is equivalent to the luteal phase levels. The half-life is approximately 30 mins and baseline concentration is reached in 12 hours, thus requiring BD dosage.
- *Medroxyprogesterone acetate*: It is completely bioavailable and highly effective after oral administration. Its half-life is 2–3 days.
- *Dydrogesterone*: It has rapid oral absorption and reaches C_{max} in 1–2 hours. It has a half-life of 4–5 hours.
- *Norethindrone/Norethisterone acetate*: It has rapid oral absorption with 65% bioavailability and has a half-life of 8 hours.
- *Levonorgestrel*: It is completely bioavailable reaching C_{max} in 1–2 hours and has a half-life of 16 hours.
- *Gestodene*: It is 100% bioavailable with a half-life of 12–18 hours.
- *Desogestrel*: It is metabolized to 3-ketodesogestrel during first pass through gut wall and liver and is 75% bioavailable. It has a half-life of 12 hours.
- *Norgestimate*: Its pharmacokinetics is not well established.

Intramuscular Administration

With intramuscular injection of 10 mg progesterone suspended in vegetable oil, maximum plasma concentrations (C_{max}) are reached at approximately 8 hours after administration, and serum levels remain above baseline for about 24 hours.[34] Doses of 10 mg, 25 mg, and 50 mg via intramuscular injection result in mean maximum serum concentrations of 7 ng/mL, 28 ng/mL, and 50 ng/mL, respectively.[34] With intramuscular injection, a dose of 25 mg results in normal luteal phase serum levels of progesterone within 8 hours, and a 100 mg dose produces mid-pregnancy levels.[41] At these doses, serum levels of progesterone remain elevated above baseline for at least 48 hours, with a half-life of about 22 hours.[8,41] Due to the high concentrations achieved, progesterone by intramuscular injection at the usual clinical dose range is able to suppress gonadotropin secretion from the pituitary gland, demonstrating antigonadotropic efficacy (and therefore suppression of gonadal sex steroid production).[34]

Vaginal or Rectal Administration

With vaginal and rectal administration, a 100 mg dose of progesterone results in peak levels at 4 hours and 8 hours after dosing respectively, with the levels achieved being in the serum luteal phase range.[41] Following peak serum concentrations, there is a gradual decline in plasma levels, and after 24 hours, serum levels typical of the follicular phase are reached. Gels (4% and 8%) and inserts are available. It should not be administered with other vaginal preparations. If therapy with another agent administered intravaginally is needed, a gap of 6 hours before or after progesterone vaginal gel is recommended.[41]

- *Progesterone 4% vaginal gel*: Each prefilled applicator delivers approximately 1.125 g of gel (45 mg of progesterone).
- *Progesterone 8% vaginal gel*: Each prefilled applicator delivers approximately 1.125 g of gel (90 mg of progesterone).
- *Progesterone vaginal insert*: Each insert has 100 mg progesterone and comes in a pack of 21 inserts with applicator.

Transdermal Administration

Transdermal progesterone is about 5-7 times stronger than oral progesterone.[32] This is due to the fact transdermal administration bypasses first-pass metabolism. As such, 20-30 mg/day transdermal progesterone is equivalent to about 100-200 mg/day oral progesterone. Some researchers have reported that absorption of progesterone via the transdermal route is poor, impractical, and unsubstantiated, however they may have been measuring serum blood or urine levels.[48,49] Other studies have shown that transdermal absorption of progesterone cream is much higher when capillary blood or saliva testing is used.[50]

Subcutaneous Injection

Progesterone can also be administered alternatively via subcutaneous injection, with the new aqueous formulation Prolutex in

Europe being intended specifically for once-daily administration by this route.[8,51,52] This formulation is rapidly absorbed and has been found to result in higher serum peak progesterone levels relative to intramuscular oil formulations.[52] In addition, subcutaneous injection of progesterone is considered to be easier, safer (less risk of injection site reactions), and less painful relative to intramuscular injection.[52] The terminal half-life of this formulation is 13 to 18 hours, which is similar to the terminal half-lives of oral micronized progesterone and intramuscular progesterone.[8]

STABILITY AND STORAGE

Oral Capsules

May be stored in tight, light-resistant containers at 25°C (may be exposed to 15–30°C).[45]

Parenteral Injection

The ideal temperature for storage is 15–30°C.[44]

Vaginal Gel and Inserts

Can be stored at 25°C (the range being 15–30°C).[46,47]

CHAPTER 5

Micronization

Micronization process by decreasing the particle size enables rapid absorption and dissolution resulting in increased bioavailability of the drug. Coating the drug with lipid (progesterone is packed in peanut oil) and dispensing it in a gelatin capsule enhances its lymphatic absorption and further increases its bioavailability. With all this, there is a predictable increase in serum progesterone concentration in a dose-dependent manner.[46]

After a single dose of 100 mg of micronized progesterone (MP), serum progesterone levels equivalent to luteal phase progesterone concentration can be achieved. For hormone replacement therapy (HRT), MP300 mg is given in two divided doses: 100 mg in the morning and 200 mg in the evening. A higher dose is given in the evening as progesterone causes some sedation as a side effect. Maximum serum levels of progesterone are higher when MP is given after breakfast as absorption is higher when taken with food. There is no adverse effect on serum lipids, rather it has an effect of reducing triglycerides. It causes mild diuresis. A 200 mg oral dose of MP approximates the antimineralocorticoid activity of 25–50 ng of spironolactone. The progesterone metabolite desoxycorticosterone has stronger mineralocorticoid activity. MP has antiestrogenic and anxiolytic effects. It has no effects on carbohydrate metabolism and blood pressure. Apart from being short-acting and requiring frequent dosing, it matches with all the favorable effects of endogenous progesterone. Intramuscular injection of the drug resulted in adequate mid-luteal serum level, and by the vaginal route the drug concentration in the endometrium is ten times higher than by the intramuscular route. There is a synchronous endometrial histology. The advantages of the vaginal route include avoidance of local pain and first pass metabolism, rapid absorption, lack of sedation, high bioavailability and reservoir effect of vagina. MP can also be administered by intranasal, sublingual and rectal routes. Its metabolite, 20α-dihydroprogesterone, has also been instrumental in eliciting progestational effects in target tissues as its peak coincides with that of the native compound.

The effects of MP include progestational, antiestrogenic and antimineralocorticoid effects.

Indications for its use include luteal phase defects (LPD), progesterone challenge test (PCT), hormone replacement therapy (HRT), endometrial protection in abnormal uterine bleeding (AUB), premenstrual syndrome (PMS) and preterm labor (PTL).

CHAPTER 6

Pharmacokinetics and Metabolism

ABSORPTION

The route of administration impacts the effects of progestins.

Progestins are rapidly absorbed with a longer half-life than progesterone and maintain stable levels in the blood.[53]

Oral micronized progesterone has a wide inter-individual variability in absorption and bioavailability. The absorption and bioavailability of oral micronized progesterone is increased approximately two-fold when it is taken with food.[54]

BIOAVAILABILITY

Progesterone is rapidly absorbed from the gastrointestinal tract and peak plasma concentrations are achieved within 2-3 hours.[45]

The peak plasma concentrations are attained in 8-9 hours following intramuscular (IM) administration.[43,47]

After intramuscular administration of 10 mg of Progesterone in oil, maximum plasma concentrations (geometric mean of 7 ng/mL) were reached within approximately 8 hours after injection and plasma concentrations remained above baseline for about 24 hours after injection. Injection of 10, 25, and 50 mg resulted in geometric mean values for maximum plasma concentration (C_{max}) of 7, 28, and 50 ng/mL, respectively.

The peak plasma concentrations are attained in 5-7 hours following administration of progesterone vaginal gel. Absorption is prolonged due to sustained-release properties of this preparation.[47]

The peak plasma are reached in 17-24 hours following administration of progesterone vaginal insert.[46]

DISTRIBUTION

Progestogens are is extensively bound to plasma proteins. Almost 96-99% of progestogens are bound, 50-54% is bound to albumin and 43-48% to cortisol-binding protein.
They are distributed in human milk.[44-47]

HALF-LIFE

Oral dose: 16-18 hours
IM doses: 20-30 hours[47]
Vaginal gel: 30-60 hours[47]
Significantly elevated serum levels of progesterone are maintained for about 12 hours.

METABOLISM

Progesterone, when taken orally, undergoes gastrointestinal and hepatic metabolism to form hydroxylated metabolites, which in turn are metabolized into sulfate and glucuronide derivatives. Progesterone is metabolized primarily in the liver by reduction. It is reduced to a variety of active and inactive metabolites, including pregnanediol, pregnanetriol, and pregnanolone (pregnanedione). Subsequent conjugation results in the formation of glucuronide and sulfate metabolites. Further metabolism may occur in the gastrointestinal tract.[46] The mean plasma metabolic clearance rate in women of reproductive age is 2510 ± 135 (SEM) L/day.

EXCRETION

The glucuronide and sulfate conjugates of pregnanediol and pregnanolone are excreted in the urine and bile. Progesterone metabolites which are excreted in the bile may undergo enterohepatic recycling or may be excreted in the feces.

Following administration of progesterone injection, 50-60% of metabolites are excreted by the kidneys, while approximately 10% are excreted by the bile and feces.[46,47] A small portion is excreted in the bile unchanged.

About 10-20% of progesterone is excreted as pregnanediol. In urine, pregnanediol glucuronide is the major metabolite of progesterone.[14] Its levels in the urine in the preovulatory phase is less than 1 mg/day, which increases to 3–6 mg/day postovulation. This level is maintained until two days prior to menses. Now a days the urine pregnanediol assay has little clinical use. It has been found to constitute about 30% of an injection of progesterone.[14]

METABOLITES

Pregnanetriol is the chief metabolite of 17α-hydroxyprogesterone. It is clinically useful to diagnose adrenogenital syndrome, in which an enzyme deficiency impairs production/synthesis of glucocorticoids and thus causes low levels of cortisol which results in elevated levels of ACTH. This causes adrenal hyperplasia which is virilizing because of production of androgens and corticoid precursors in abnormal amounts. The accumulation of 17α-hydroxyprogesterone leads to an increased excretion of pregnanetriol in urine. The serum or plasma assay of 17α-hydroxyprogesterone is more accurate diagnostic method than urinary pregnanetriol measurement. The range for normal levels of 17α-hydroxyprogesterone are <100–200 ng/dL. The levels are elevated 10–400 times in congenital adrenal hyperplasia.

Progesterone is reduced to its much less potent progestogen metabolite 20α-hydroxyprogesterone by the 20α-hydroxysteroid dehydrogenases AKR1C1 and AKR1C3. In addition, progesterone is converted into its active progestogen and neurosteroid metabolite 5α-dihydroprogesterone by 5α-reductase, which in turn can be converted into the non-progestogen but even more potent neurosteroid allopregnanolone by 3α-hydroxysteroid dehydrogenase.[55] Progesterone is converted into 17α-hydroxyprogesterone and deoxycorticosterone by 17α-hydroxylase and 21-hydroxylase, respectively. 17α-hydroxyprogesterone serves as a precursor for the androgens and estrogens, and deoxycorticosterone is a corticosteroid and a precursor for other corticosteroids as well.

Neurosteroid Metabolites

A portion of progesterone is converted into 5α-dihydroprogesterone and allopregnanolone (a conversion that is catalyzed by the enzymes 5α-reductase and 3α-hydroxysteroid dehydrogenase [3α-HSD] and occurs in the liver, reproductive endocrine tissues, skin, and the brain), which are neurosteroids and potent potentiators of GABA receptors.[55-57] It is for this reason that common reported side effects of progesterone include dizziness, drowsiness or sedation, sleepiness, and fatigue, especially at high doses.[56,57] As a result, some physicians may instruct their patients to take their progesterone before bed. Both oral and intramuscularly injected progesterone produce sedative effects, indicating that first-pass metabolism in the liver is not essential for the conversion to take place.[58-60] Moreover, the sedative effects occur in both men and women, indicating a lack of sex-specificity of the effects.

SPECIFIC METABOLISM

- *Medroxy progesterone acetate*: It is metabolized like progesterone
- *Levonorgestrel and norethindrone*: They also undergo reduction and hydroxylation followed by formation of conjugates of which sulfates are predominant which also contribute to progestational activity.
- *Desogestrel*: It is a pro-drug which is converted to active compound 3-ketodesogestrel in the liver.
- *Gestodene*: It is not a pro-drug and is excreted in urine as conjugate.
- *Norgestimate*: It is partly a pro-drug for levonorgestrel.

CHAPTER 7

Physiological Actions of Progesterone

Progesterone acts through its receptors. The luteinizing hormone (LH) stimulates the receptor expression and receptors begin to appear in the granulosa cells of the dominant follicle in the periovulatory period. Progesterone receptor expression varies in different phases of the menstrual cycle. In the late proliferative and early secretive phase the receptor expression in the endometrial glandular epithelium is the maximum as induced by estrogen and by midpoint of secretory phase the expression reaches the lowest. Endometrial stromal cells do not show any variation in the receptor expression. The receptors are absent in the decidual epithelial cells, but strongly present in the decidualizing stromal cells. Smooth muscle cells of the uterus show a strong receptor expression throughout the cycle. Progestins cause a decrease in the estrogen receptor levels, affect the enzymes leading to excretion of estrogen from the cells and suppress estrogen-mediated transcription of oncogenes. Progestins are powerful anti-estrogens.

Low levels of progesterone in the presence of estrogen cause the midcycle FSH surge and rising levels of LH. The LH surge causes ovulation of the follicle. The LH causes luteinization in the ovulating follicle and its granulosa layer starts to secrete progesterone into the blood. The presence of the oocyte inhibits luteinization, hence progesterone secretion is suppressed and very low levels reach the brain. After ovulation, rapid and full luteinization leads to a marked increase in progesterone levels, which along with estrogen suppress gonadotropin secretion by a negative feedback mechanism.

Progesterone has an important role in mediating the slowing of GnRH pulses in the late luteal phase which causes a rise in FSH necessary for initiating the next cycle.

High levels of progesterone inhibit ovulation at the hypothalamic level, whereas low levels of progesterone exert a facilitatory action only at the pituitary level in response to GnRH. Progesterone affects the positive feedback response to estrogen in a dose and time dependent manner. After adequate estrogen priming, it facilitates the response directly acting on the pituitary and causing LH surge in the presence of subthreshold levels of estrogen. This is the reason why some anovulatory amenorrhoeic women ovulate after a progesterone challenge test. However, it blocks the LH surge if given before estrogen stimulus or in high doses.

During pregnancy, progesterone is formed predominantly in the placenta. The fetal and the maternal work as a unit for steroidogenesis, with the placenta using precursors from either of them. The maternal compartment of the placenta obtain cholesterol and pregnenolone from the bloodstream for progesterone synthesis. The fetal contribution is negligible as is evidenced by the presence of high progesterone levels after fetal demise.

Progesterone prepares and maintains the endometrium to allow implantation of the fertilized zygote. Progesterone is important in preventing maternal rejection of trophoblast by suppressing the maternal immunologic response to fetal antigens. Until 10 weeks of gestation, progesterone is largely produced by the corpus luteum with the pregnancy solely depending on it till the seventh week. The predictive value of progesterone measurements is limited as individual variation is great. Between the seventh and the tenth week of gestation, the corpus luteum and the placenta share the synthesis of progesterone. Production gradually increases as the pregnancy advances, with levels of 10 ng/mL rising to 100–200 ng/mL at term. At term, placenta produces about 250 mg/day of progesterone, derived mainly from cholesterol and most of which enters the maternal circulation. This production is independent of the precursor amount available, fetal well-being, a live fetus and uteroplacental perfusion.

Estrogen increases the process of endocytosis involving the low density lipoprotein (LDL) cell membrane receptors during pregnancy which helps in utilizing the LDL-cholesterol in the trophoblast. Amino acids are formed following hydrolysis of the protein component of LDL and they are utilized by the fetus.

Progesterone is also synthesized by human decidua and fetal membranes where pregnenolone is the most important precursor and not cholesterol. This local steroidogenesis plays an important role in parturition regulation. Progesterone withdrawal, by its increased metabolism to 17α-hydroxyprogesterone, is associated with a decrease in the resting potential of myometrium, i.e. an increased response to oxytocic stimuli. Dihydroxyprogesterone is a precursor for estrogen which rises a few days prior to parturition.

Estrogens increase the vascularity and permeability and rhythmic contractions, thus estrogen rise coupled with progesterone withdrawal result in enhancement of conduction and excitation. These events are secondary to direct induction of a placental enzyme by fetal cortisol. The clinical application is seen in patients of preterm labor administered progesterone pharmacologically.

The progesterone receptor concentration in the myometrium does not change during advancing pregnancy and at the onset of labor. But a shift in the receptor isoforms occurs with a dominance of progesterone receptor A and other isoforms

SUMMARY OF ACTIONS OF PROGESTERONE

- It transforms proliferative endometrium into secretory endometrium. It acts on the endometrium, decreases estrogen receptors and mitotic activity and develops the stromal component of the endometrium. It thus prepares the endometrium for future prostaglandin production, permitting complete uniform shedding at the time of progesterone withdrawal.
- It is essential for implantation of the ovum and for maintenance of pregnancy.
- It stimulates growth of mammary alveolar tissue and promotes mammary gland development.
- It relaxes uterine smooth muscle.
- It has very minimal estrogenic and androgenic activity.
- When used as part of an assisted reproductive technique program in the luteal phase, progesterone supports embryo implantation.
- It inhibits (at the usual dose range) the secretion of pituitary gonadotropins, which in turn prevents follicular maturation and ovulation.
- In larger doses, it decreases the effect of estrogen, which includes epithelial stimulation of endometrium, breast and metabolic adversities. This effect plays a major role in preventive oncology in gynecological—uterine/endometrial cancer and breast cancer and prevention of systemic disease such as cardiovascular risks, osteopenia and improved function of nervous system.

CHAPTER 8

Pharmacodynamics and Selectivity of Progestins

Pharmacologic effects of progestins that are clinically significant are as follows (Table 8.1):

1. **Suppression of hypothalamic-pituitary axis:** The preovulatory surge of gonadotropins is suppressed at the level of hypothalamus and pituitary, thus forming the basis of their contraceptive use.
2. **Effects on endometrium:**
   Luteal phase of menstrual cycle:
   - Stimulate secretory activity in glands
   - Induce stromal edema
   - Predecidualization of stromal cells
   - Coiling of blood vessels.

   Peri-implantation period:
   - Decidualization of stroma
   - Hypersecretory activity
   - Modulation of prostaglandin activity
   - Stimulate prolactin secretion from endometrial cells
   - Expression of integrins subunits α-4 and α-3.

   These effects are progestational and prepare the endometrium for reception and support of pregnancy/fertilized zygote/ovum.
3. **Effects on myometrium:**
   - Decrease frequency and amplitude of myometrial contractions
   - Inhibit formation of gap junctions
   - Modulate prostaglandin and oxytocin action.
4. **Effects on lipid metabolism:** Progestins have anti-estrogenic and androgenic action cause:
   - Decrease in HDL (high-density lipoprotein) cholesterol levels
   - Decrease in triglycerides
   - Increase or no change in LDL (low-density lipoprotein) cholesterol.

Table 8.1: Relative pharmacodynamics of progestins.

	Inhibition of ovulation	Transformation of endometrium	Binding to progesterone receptor (%)	Selectivity ratio (Prog: androg. receptor binding)	↓ SHBG	HDL ↓
Progesterone	+	++	100	+++	±	±
Medroxyprogesterone acetate	+	++	−100	+++	±	+
Norethindrone	++	++	305	+	+	++
Norgestrel	+++	+++	628	++	+	++
Desogestrel	+++	+++	400–500	+++	±	±
Gestodene	+++	+++	11 (metabolites—628)	+++	−	±
Norgestimate	++	+++	?	+++	−	?

(SHBG: Sex hormone binding globulin; HDL: High-density lipoprotein).

These changes are proatherogenic and negate the benefits of estrogen in hormone replacement therapy on lipid metabolism. They are more prominent with 19-norsteroids viz. norethindrone, but are least with the newer progestins.

5. **Androgenic actions:** Progestins have varying degree of androgenic effects by:
   • Direct action on testosterone receptor
   • Decrease in sex hormone binding globulin (SHBG) levels.

   These are responsible for their side effects like acne, hirsutism, etc. and these are more marked with 19-norsteroids. Natural progesterones and newer progestins have virtually no androgenic effects.

6. **Effects on carbohydrate metabolism:** Variable findings have been observed in this regard in various studies. A few have reported abnormal glucose tolerance test in 5–15%, but most

studies have found no significant change in carbohydrate metabolism by any type of progestin.

7. **Effects on hemostatic factors:** Progestins alone have not been associated with any alterations in hemostatic factors. Various studies on desogestrel, which was once thought to be more atherogenic, have refuted any such effect. However, in predisposed women progestins can contribute to development of thrombosis by increasing the distensibility and capacitance of the veins.
8. **Effects on nervous system:** Progestogens have been found to have effects on cognitive functions, mood, feeling of well-being and on dementia of Alzheimer's disease.
9. **Effects on bones:** Studies have found that combined estrogen and progesterone therapy is more effective in osteoporosis than estrogen replacement therapy. Animal studies have shown that progesterone inhibits the decrease in bone resorption parameters, whereas the bone formation parameters remain elevated.

CHAPTER 9

Individual Progestogen Description

DYDROGESTERONE

Of all the progestins, it has properties closest to native progesterone. It is also known as retroprogesterone or stereoisomer of progesterone as the native molecule is modified stereochemically rather than by attachment of side chains.

It has no estrogenic, androgenic or anti-ovulatory effects. It is absorbed rapidly by oral route and peak plasma levels are seen within 20 minutes. The daily dose ranges from 10 to 20 mg. The biological effects include decidualization of endometrium and it is 10-15 times more potent than progesterone. The safety profile is comparable. Complete secretory transformation is possible by a dose of 10 mg/day for 10 days, with no glandular stromal asynchrony and thus it can cause predictable withdrawal bleeding in an estrogen primed endometrium within 3-5 days of cessation of therapy.

Metabolically there is no effect on body weight, blood pressure, blood clotting factors and lipoproteins. It is a diuretic and prevents sodium retention. Adrenal and renal functions are not affected.

Clinically it is used in progesterone challenge test, luteal phase defect, sequential regimens of hormone replacement therapy and abnormal uterine bleeding.

Recent data regarding the immunological basis of abortion suggest there is a decrease in progesterone induced blocking factor (PIBF) in patients with recurrent abortions and threatened abortion. PIBF changes the balance of immune system mediators such as cytokines. There are two types of cytokines—T-helper 1 (Th1) and T-helper 2 (Th2). Th1 are harmful and Th2 are helpful with regards to pregnancy. PIBF leads to decreased Th1/Th2 ratio. Dydrogesterone effectively induces PIBF production thereby decreasing Th1 cytokines. In 86 out of 146 women with threatened abortion, 75% who received dydrogesterone experienced a successful outcome of pregnancy, whereas only 66% of women without any therapy had a successful outcome.

ESTERS OF PROGESTERONE

Addition of hydroxyl group at 17 position yields a compound which is inactive due to rapid excretion or transformation, but its ester derivative has progestational activity, which is used parenterally due to its first pass hepatic metabolism.

Substitution of methyl group at 6th position yields 6-methyl acetoxyprogesterone or medroxyprogesterone acetate (MPA). It has similar action like the parent compound, but more selective action with lesser endometrial stromal asynchrony in development of secretory endometrium. It has no estrogenic or anti-estrogenic activity. Being an ester it is long-acting. It is efficacious both enterally and parenterally. Its absorption is rapid and half-life is 8 hours, so a single daily dose is needed. For most women, 10 mg for 12 days each month is adequate for endometrial protection. Due to prolonged hypothalamic-pituitary axis suppression, it is not useful in sequential regimes of hormone replacement therapy (HRT) where withdrawal bleeding is warranted. However, this property is advantageous for prolonged contraception and in preventive oncology, i.e. primary and secondary endometrial carcinoma.

Medroxyprogesterone acetate has been shown to have anti-tumor activity in endometrial and breast carcinoma. In breast carcinoma, its anti-tumor activity shows two levels of therapeutic efficiency. The response rate with low dose is 25% and with high dose is 40%. It has anti-cachectic effect by reducing IL-6 secretion from tumor cells. The therapeutic efficacy of medroxyprogesterone acetate as an injectable is well known. It is used worldwide in 9 million women of reproductive age group. It can also be used for endometrial hyperplasia, but it is not as effective as norethisterone. The other uses are endometriosis, progesterone challenge test and in combined regimens of HRT in low doses. Like the other synthetic progestins, MPA also lowers HDL, but to a lesser degree than the androgenic steroids.

The derivatives with a 17α-acetoxy group are cyproterone acetate and chlormadinone.

Depot Medroxyprogesterone Acetate (DMPA)

It was the first popular long-acting progestin-only contraceptive. It is a microcrystalline suspension of low solubility which provides a sustained release of the hormone when given by the intramuscular route. Blood levels of the hormone peak within 24 hours and decline over a period of more than 90 days. It is advisable to administer the injection in a big muscle like the gluteal muscle. The injection is given deep intramuscularly in a Z fashion. The injection vial should be thoroughly shaken before administration. The site of injection should not be massaged after the injection. The dose is 150 mg every three months. In some parts of the world a dose of 500 mg every 6 months was tried with fair results. The contraceptive effect of DMPA is due to powerful progestational effect seen in the form of thickened cervical mucus, atrophic endometrium and a decrease of tubal motility. All these factors impede sperm travel/migration in the reproductive tract. It does not have any estrogenic or anti-estrogenic activity.

It is initially given on day 1-7 of the cycle. The contraceptive effectiveness continues for a few weeks after the due date of the next injection. The life-time failure rate is $\leq 0.1/1000$. There is no absolute contraindication other than pregnancy and coagulation disorder. It is one of the long acting reversible contraceptive (LARC) method.

Subcutaneous DMPA

Subcutaneous (DMPA) is an innovative method which makes injections simpler. It is 99% effective at preventing unintended pregnancy when given correctly and on time every three months and is discreet contraception for women and adolescent girls. It comes as a prefilled and ready to inject syringe and has lower dose of DMPA (104 mg) with a 2.5-centimeter needle. It is possible for women to self-inject with proper training. Product is registered for self-injection in the United Kingdom, several European countries, and in an increasing number of Family Planning (FP) 2020 countries. Self-injection is also supported by the World Health Organization. Based on its lower dose, DMPA-SC is expected to have a side-effect profile that is similar to or better than that of DMPA-IM.

The major advantages are its safe use in the postpartum period and in women with menorrhagia. It causes relief in premenstrual syndrome and a decrease in sickling crises. It offers other advantages like reduction in pelvic inflammatory disease, anemia and dysmenorrhea.

As with other hormonal contraceptives, the menstrual cycles are disturbed and there are irregular bleeding episodes, long cycles, scanty bleeding or amenorrhea. This is due to increased vascular fragility and thinning of the vessel wall in endometrium along with local molecular interactions. The menstrual irregularities are more often seen in lactating women than in post-abortal women. Minor side effects include weight gain, nausea, headache and bloating. Fertility is often delayed after cessation of DMPA injections, so women who wanted predictable return of fertility upon cessation of contraception should not be advised DMPA. The median conception time is 5–7 months.

The carcinogenic effect of DMPA seen in Beagle dogs has not been observed in humans. Rather it has a protective effect against endometrial and breast carcinoma. The contraindications for its use are undiagnosed uterine bleeding, breast pathology, coagulation disorders, and women who want faster return of their fertility after contraception.

ESTRANES (19-NORTESTOSTERONE DERIVATIVES)

Estranes are derived from 19-nortestosterone, the parent compound used in oral contraceptives. They are characterized by the presence of an ethinyl group at position 17α and by the absence of methyl group between A-B rings. Most of the estrane progestogens resembling norethindrone or norethisterone are converted to more potent compounds as norethynodrel (isomer) and others. This compound combined with ethinyl estradiol has a long track record of efficacy and relative safety as oral contraceptive.

Norethindrone is also used as progestin-only pill (POP) in a dose of 350 µg/day. It is used as a hemostatic agent in anovulatory irregular bleeding in a dose of 20–30 mg/day in divided doses and as maintenance therapy during the cycle for 3–6 months. The contraceptive effect of these is due to a strong inhibitory action on pituitary gonadotropins. They also cause endometrial proliferation of a type that differs from the luteal phase of the normal cycle.

GONANES

In the last decade, newer progestins with lesser adverse effects on lipid metabolism belonging to the category of gonanes were added. They are derivatives of 19-nortestosterone. The possibility of using these progestins in HRT formulations has also been explored.

The following compounds belong to this class of progestins:

Desogestrel

This is 13-ethyl-11-methylene-18,19-dinor-17-alpha-preg-4-en-20-yn-17 in structure. It was introduced in the US in the need to decrease the side effects and undesirable metabolic changes in a bid to make the oral contraceptives more desirable and usable to more women. It is similar to levonorgestrel in chemical structure but acts through the principle metabolite, 3-ketodesogestrel. Only a minimal action occurs via levonorgestrel.

It has lesser androgenic activity, lesser effects on carbohydrate metabolism as compared to levonorgestrel. It has no adverse effects on lipid metabolism and rather has a positive effect on HDL. Its progestational effects like inhibition of ovulation, delay in menses, transformation of endometrium and binding to progesterone receptors at a very low dose level, has evolved it into a progestinonly pill. Its half-life is 38 hours and a steady serum level is achieved in 8–10 days. A dose as low as 60 μg can exert a contraceptive effect. In monophasic pills, its dose is 150 μg, and in triphasic pills, it is used in successive doses of 50, 100 and 150 μg per week. The cycle control is adequate and is better in triphasic formulation when compared to triphasic norethindrone. Failure rate is less than 1% per year. It decreases androgen related side effects by increasing the sex hormone-binding globulin (SHBG) levels. Desogestrel with low doses of estradiol, 20 μg, has emerged as a better method for perimenopausal contraception.

Gestodene (GSD)

It is the lowest dosed and most potent amongst the newer progestins. It is a delta-15-levonorgestrel and is best characterized as 13-ethyl gonane. The bioavailability with a dose of 75 μg is 100%. This is due to slow metabolism and decreased activity of P-450 monooxygenase and 5α-reductase involved in the inactivation of GSD. The progestogenic and anti-estrogenic activity is approximately twice that of levonorgestrel (LNG), but androgenic activity is the same. Thus the ratio of progestogenic to androgenic activity is higher than that of LNG. The high contraceptive efficacy is based on the prevention of ovulation and marked progestogenic and anti-estrogenic effects on cervical mucus and the endometrium.

In monophasic pills, GSD is used in the dose of 75 μg and in triphasic pills, it is used in the successive doses of 50 μg (6 days), 70 μg (5 days) and 100 μg (10 days). In GSD containing pills, there is no need to use another method even if two doses are missed, unlike the usual oral contraceptive pill (OCPs) GSD results in a good cycle control, especially with the triphasic pills. Because of a weak androgenic effect of GSD, these contraceptive formulations are characterized as estrogen dominant with respect to their hepatic effects. There is minimal rise in triglycerides, increase in SHBG and decrease in androgenic side effects. There is a slight impairment of carbohydrate metabolism in predisposed women and a slight increase in thromboembolic disease, but both effects are not clinically significant. There is no change in blood pressure. The drug continuation rates are more than 90%.

Norgestimate (NGM)

This is also a 19-nortestosterone derivative of the gonane family. It is also a prohormone. Its metabolic activity results mainly through 17-deacetyl-norgestimate and some activity is through levonorgestrel. NGM containing OCPs, either as monophasic or as triphasic pills perform as well as those containing the older progestins. The androgenic side effects are lesser than LNG, but more than the newer progestins. NGM formulations have not found to alter carbohydrate metabolism over a period of use for 24 months. It increases SHBG levels and does not cause weight gain or increase in blood pressure. More long term studies are needed for comparative evaluation with other progestins.

NEWER PROGESTINS

Nomegestrol

It is derived from progesterone after elimination of the 19 carbon and is one of the most potent progestins. Its progestational, antiandrogenic and ovulation inhibiting activity allows it to be used in various indications. It is used in Europe for postmenopausal HRT and contraception. The first 24 pills contain 2.5 mg of nomegestrol acetate and 1.5 mg estradiol and the last four are placebos. It has a potent inhibitory effect on gonadotropins, no androgenic activity and rather antiandrogenic activity. Unlike drospirenone, it has no anti-mineralocorticoid or glucocorticoid activity. It does not cause sodium retention nor does it has anti-diuretic activity. It is used in 1.25 mg daily dose. No major side effects have been noted. In some clinical trials an increase in irregular bleeding was noted and this may be due to its relative lack of endometrial effects. It has a much longer half-life than MPA. In a 2-year study using an implant with 55 mg nomegestrol, no effects on SHBG and carbohydrate and lipid metabolism were observed.

Individual Progestogen Description 33

Dienogest

It is a 19-nortestosterone having a cyanomethyl group instead of an ethinyl group in the 17 position and an additional double bond. It has properties of the derivatives of progesterone and 19-nortestosterne family. It has anti-androgenic activity and exhibits a weak binding affinity for progesterone receptors but a profound activity on the endometrium. It has an effect on estrogen and other steroids. It is five times more potent than LNG and ten times more potent than MPA. It inhibits ovulation by peripheral action. Its anti-androgenic activity is 30–70% of cyproterone acetate. In Germany, a combined oral contraceptive (COC) containing 2 mg of dienogest and 30 µg of ethinyl estradiol (EE) has been available since 1991. Various studies have shown no major side effects or metabolic derangements. It is incorporated into a unique 4-phase contraceptive pill with estradiol valerate decreasing from 3 mg to 1 mg over 26 days and dienogest 2 mg on days 3–7 and 3 mg on days 8–24. This has shown comparable bleeding profile as the low-dose monophasic pill. Its action on the endometrium is depicted in the treatment of endometriosis.

Nestorone

It was formerly known as ST-1435. Its progestational activity is higher than LNG. It has only 10% bioavailability. Single rod subdermal implants with a life span of 2 years have been developed and have proved to be highly effective contraceptives. It does not alter liver function, carbohydrate or lipid metabolism. It has the highest incidence of oligomenorrhea and amenorrhea. It has great potential for use in lactating women. It is also being studied in the form of vaginal rings, transdermal gels and patches.

Drospirenone

It is an analog of spironolactone having high affinity for mineralocorticoid receptors that produces an antimineralocorticoid activity and its biochemical profile matches that of progesterone. This results in a rise in plasma aldosterone and renin as compensatory effect of drospirenone. A COC containing 30 µg of EE and 3 mg of drospirenone (Yasmin) resulted in natriuretic effect, weight loss and good cycle control. Drospirenone increases the SHBG levels thereby decreasing androgen levels. Caution is advisable regarding serum potassium levels, though clinical hyperkalemia is not seen usually. Its use is best avoided in women with abnormal renal, adrenal or hepatic function. The therapeutic role of drospirenone in premenstrual syndrome or premenstrual dysphoric disorder (PMDD) has been studied and the beneficial impact is minimal after two years use, specifically in acne, appetite and food cravings. Yaz, a 24-day version of this pill uses 20 µg of EE.

Trimegestone

It is still under development. It has anti-androgenic activity. The progestational effect is greater than MPA and it has anti-mineralocorticoid activity in higher doses.

The relative activity of all progestins is shown in Table 9.1.

Table 9.1: Relative activity of various progestins.

Drug	Progestogenic action	Androgenic action	Anti-androgenic action	Antimineralocorticoid action	SHBG
Progesterone	1	–	+	+	–
Cyproterone acetate	4	–	+++	–	–
Norethisterone	4	+	–	–	–
MPA	4	+	–	–	–
LNG	6	++	–	–	++
Desogestrel	8	+	–	–	–
Gestodene	9	+	–	+	+
Norgestimate	4	+	–	–	–
Drospirenone	4	–	+	+	+
Nomegestrol	5	–	+	–	–
Nestorone	10	–	–	–	–
Trimegestone	10	–	+	+	?
Dienogest	4	–	+	–	–

(SHBG: Sex hormone-binding globulin; MPA: Medroxyprogesterone acetate; LBG: Levonorgestrel).

CHAPTER 10

Clinical Usage Guidelines of Progestins

The uses of progestins can be divided as diagnostic and therapeutic as shown in Table 10.1.

DIAGNOSTIC USE

Progesterone challenge test (PCT): This test is used in patients with amenorrhea, whether primary amenorrhea, secondary amenorrhea, lactational amenorrhea or amenorrhea in perimenopausal period. For this purpose, short course of a progestogen is given. A withdrawal bleeding after 3-5 days of progesterone cessation indicates an estrogen-primed endometrium. If used before ovulation induction, natural progesterone is to be preferred over the synthetic progestins as the latter suppresses the hypothalamus. The following regimes can be used:
- Micronized progesterone: 300 mg once a day for 3-5 days
- Dydrogesterone: 10 mg once a day for 10 days

Table 10.1: Therapeutic applications of progestins.

Gynecological indications	Family planning/Contraception indications	Obstetric indications
Infertility and LPD	OCPs	PTL
ART	POPs	Threatened and recurrent abortions
AUB	Implants	
Endometriosis	Injectables	
Myomas	IUCDs	
PMS	Vaginal rings	Postpartum depression
HRT		
Amenorrhea		
Endometrial hyperplasia		
Endometrial cancer		
Postponement of menses/ Self-regulated menses		

(LPD: Luteal phase defect; ART: Assisted reproductive technology; AUB: Abnormal uterine bleeding; PMS: Premenstrual syndrome; HRT: Hormone replacement therapy; OCP: Oral contraceptive pill; POP: Progestogen-only pill, IUCD: Intrauterine contraceptive device, PTL: Preterm labor).

- Medroxyprogesterone acetate: 10 mg once a day for 10 days
- Progesterone (in oil) injection: 100 mg single dose is administered
- Progesterone 4% vaginal gel: One prefilled applicator every other day for a total of 6 doses.

THERAPEUTIC USES

Contraceptives

Progestogens are used as contraceptives, either alone or in conjunction with estrogens. Progestogens act as contraceptives by various mechanisms. They aid estrogens in their anti-ovulatory effect. They make the cervical mucus thick and hostile for sperm migration, reduce the tubal motility. They make the endometrium unfavorable for implantation. They are used as combined oral contraceptive pills (OCPs), progestogen-only pills (POPs), injectable contraceptives, subdermal implants and progestin containing intrauterine devices (IUDs).

WHO Medical Eligibility Criteria for progestogen containing contraceptives.

Contraindications to Progestogen-only Pills (WHO Medical Eligibility Criteria)

WHO 4:

Absolute contraindication	Current carcinoma, breast
WHO 3 Method of last choice	Breastfeeding <6 weeks; current deep vein Thrombosis, pulmonary embolism, past carcinoma breast with no evidence of disease for 5 years; severe cirrhosis; liver tumors; current and history of IHD or stroke; use of rifampicin and certain anticonvulsants, SLE with antiphospholipid antibodies
WHO 2	Past ectopic pregnancy, multiple risk factors for arterial cardiovascular disease; hypertension with BP >160/100 mm Hg; vascular disease; history of DVT/PE; major surgery with prolonged immobilization; hyperlipidemia; migraine with aura; unexplained vaginal bleeding before evaluation; undiagnosed breast mass; irregular bleeding; diabetes; gallbladder disease; mild cirrhosis; concurrent use of griseofulvin and antiretroviral therapy using NNRTIs.

Combined Oral Contraceptive Pills

In combined OCPs, first generation (e.g. norethisterone), second generation (e.g. levonorgestrel) and third generation (e.g. desogestrel) progestins are used. The newer progestins have little or no androgenic side effects. The selectivity index, i.e. the ratio of desired pharmacological effects to the undesired adverse effects, of desogestrel is 100, of gestodene is 60, of norgestrel is 30 and of norethisterone is 20.

The first generation OCPs contain 50 µg or more of ethynyl estradiol. The low dose OCPs contain less than 50 µg EE. The second generation OCPs contain 20, 30 or 35 µg EE with levonorgestrel, norgestimate or other members of the norethindrone family. The third generation OCPs contain 20, 30 or 35 µg of EE plus desogestrel or gestodene. The fourth generation OCPs contain drospirenone, dienogest or nomegestrol acetate.

All progestins derived from 19-nortestosterone increase insulin resistance and decrease glucose tolerance. This effect is minimal with low dose progestins and negligible with the newer progestins, and hence of no clinical significance.

The new progestins do not adversely affect the LDL-cholesterol profile due to their reduced androgenicity. Rather the estrogen progestin combination with new progestin promotes favorable lipid changes. Thus they offer clinical protection against cardiovascular diseases.

The multiphasic preparations aim for lesser metabolic effects and minimizing breakthrough bleeding by altering the dose of hormones in the cycle. But metabolism studies show no significant improvement over the low dose monophasic pills.

The estrophasic approach involves a gradually increasing dose of estrogen with a continuous low dose of progestin. This results in lesser side effects with estrogen-like nausea and is very effective for treating acne as it results in a marked increase in SHBG (sex hormone-binding globulin) levels which produce a very low androgenic state.

Vaginal Estrogen-progestin Contraception

Vaginal contraception has gained a lot of popularity in recent times owing to an increased ease of use and avoidance of a daily regimen resulting in increased compliance. The vagina acts as an excellent reservoir for drug absorption and storage plus it has the advantage of first pass effect due to close proximity to the uterus. The uncornified stratified squamous vaginal epithelium allows easier penetration of the drugs to the underlying lamina propria rich in blood vessels and the areolar connective tissue layer rich in vascular plexus. Unlike the skin, there are no fat cells, hair follicles or glands which can interfere with drug absorption. The unpredictable absorption and serum level fluctuations associated with oral route owing to vomiting, drug-drug interference or decreased intestinal absorption plus hepatic metabolism (estrogen is 95% metabolized by the liver) are practically avoided or much lowered by the vaginal route, as also with the transdermal, intrauterine and intramuscular route.

Vaginal rings include seven combination rings and six progestin-only rings. In short-acting rings for one week use, weak progestins like medroxyprogesterone acetate and progesterone are used. The more potent levonorgestrel and nesterone are used in long-acting rings for up to one year use.

The **NuvaRing** is a soft, flexible transparent ring made of 2 micron thick ethylene vinyl acetate copolymer in which are contained crystals of etonogestrel or 3-ketodesogestrel and EE. The ring is 4 mm thick and 54 mm in diameter and available in only one size that fits all women. It delivers 15 μg EE and 120 μg of etonogestrel per day. Target levels of both the hormones are reached within 24 hours of insertion and they remain stable for 35 days. The ring is worn for three weeks and then changed after withdrawal bleeding or can be worn continuously for 5 weeks and then changed. Breakthrough bleeding is effectively managed by a 4-day hormone-free interval. The ring produces circulating levels which are 30–40% of the peak levels produced by oral contraceptives, thus reducing systemic exposure and they are effective in inhibiting ovulation with a clinical pregnancy rate of <1% in trials. If the ring is not replaced within 3 hours, a backup method should be used until the ring has been in place for 7 days. It is not required to place the ring in a special position, it needs to be only in touch with the vaginal epithelium. Infections and abnormalities of the vagina are relative contraindications for its use. In the unlikely event of damage to the ring, leakage of the hormones does not occur because of the ethylene membrane of the ring. The common reasons of discontinuation are—coital difficulties, foreign body sensation, expulsion. Only 2–3% of women experience expulsion in one year of use. Removal is not needed for sexual intercourse, but if required it can be removed and replaced within 3 hours. It does not affect cervical cytology and the vaginal flora. One well-done study reported increased wetness in vagina due perhaps to an increase in lactobacilli. Neither vaginal pessaries or creams for fungal infections, nor tampon use and spermicides like nonoxynol-9 affect absorption of steroids from the ring, as proven by studies. Vaginal administration is associated with the lowest estrogen exposure as compared to with transdermal and oral methods. This explains the low incidence of estrogenic side effects like nausea, dysmenorrhea and breast tenderness. In a randomized comparison, more women preferred vaginal contraception due to this reason. Breakthrough spotting rates are much lower, there are no changes in LDL and HDL levels and the clotting parameters and insulin sensitivity are not affected. However, the SHBG and triglyceride levels are substantially increased.

A vaginal ring delivers 15 µg EE and 150 µg of nesterone daily and is effective for one year can be removed periodically to induce withdrawal bleeding like menses. Progestin only rings are useful in lactating women.

Summary of advantages of vaginal rings:
1. Elimination of a daily or coital regimen, thus better compliance
2. Avoidance of gastrointestinal absorption problems
3. Avoidance of first-pass liver effects
4. Forgiving of delays, contraceptive efficacy for 5 weeks
5. Lower systemic estrogen exposure and side effects
6. Less frequent breakthrough bleeding and spotting.

Transdermal Estrogen-progestin Contraception

The transdermal patches achieve similar efficacy as the oral contraception in clinical trials and similar spectrum of actions and side effects, but the kinetics are not identical and daily fluctuations are avoided. The weekly schedule with them is simpler than daily oral contraception with better compliance and is also less susceptible to delays and omissions. This has often resulted in overall lower pregnancy rates as compared to oral contraception.

The transdermal patch Ortho-Evra contains 600–750 µg EE with 6 mg of norelgestromin and delivers 20 µg EE and 150 µg norelgestromin daily. Norelgestromin was previously known as 17-deacetylnorgestimate and it is the primary active metabolite of norgestimate. Its resulting metabolite after liver metabolism is levonorgestrel. The patch with an area of 20 cm^2 has three layers in a matrix arrangement. The inner layer is a polyester liner which is removed from the second adhesive layer just before application. The middle layer contains the adhesive and the hormones and the outer polyester layer provides support to the middle layer. It is applied in discrete locations such as lower abdomen, inner arm, upper outer arm, upper torso, buttock. It is applied on the same day, but not on the same exact site, once each week for 3 weeks, followed by a week without any patch, if withdrawal bleeding is desired, otherwise used continuously. Timing need not be precise. Instructions for the patch are similar to those for oral contraception for first-day starts, Sunday starts or same-day starts, including the need for a backup method for 7 days unless it's a first-day start. Contact with tight clothing is avoided. The underlying skin has to be clean, dry and free of creams, lotions and irritation. After applying the patch, press on it for a few seconds to ensure its edges stick and lightly dust around the edges with talcum powder to prevent formation of lint ring. Daily activities like bathing, sauna, swimming and exercise do not cause detachment. Precautions include a watch over the patch and if it seems loose or has been partially or totally off for less than 24 hours, the same patch can be reapplied. Single extra patches are provided with the regular patch. If more than 24 hours, then a new patch is applied initiating a new cycle and new change day, with a backup method used for 7 days. Within the patch cycle, a delay of up to 2 days carries no risk, but more than 2 days requires initiation of a new cycle and change day with backup. In about 5% cases detachment occurs and of this half occur in the first cycle of use. In a study of one year, about 2–5% of the patches were replaced. Skin irritation is seen at the application site in 20% users and about 2% discontinue due to this reason. Breast discomfort is experienced by approximately 20% users and is usually not severe and only 1% discontinue because of this factor. Gonadotropin levels return to baseline by 6 weeks after discontinuation. The failure rate is less than 1%. In overweight women, extended regimen or continuous use is recommended as there is evidence that hormonal contraceptive failure is increased in overweight women (over 155 lb).

Summary of advantages with transdermal patches:
1. Better continuation rates with weekly regimen
2. Avoidance of gastrointestinal absorption problems
3. Avoidance of first-pass liver effects
4. Non-contraceptive benefits like cycle regulation.

Progestogen-only Contraceptives (Table 10.2)

These are a good option for breastfeeding women as they have no effect on lactation and milk production. There are pills, implants, intrauterine contraceptive devices (IUCDs) and injectables available for contraception. Subdermal implants available containing progestins are—Norplant-6, which consists of six rods each containing 36 mg of levonorgestrel and lasts for 5 years; Norplant II, consisting of two rods, each having 70 mg of levonorgestrel and lasting three years; Implanon, which is a single rod containing 67 mg of 3-ketodesogestrel and lasts for three years. Mirena is an IUCD containing levonorgestrel. Long-acting reversible methods of contraception (LARCs) are defined as those requiring administration less than once a month. LARC methods include levonorgestrel intrauterine system, injectable and implantable progestogens. LARC methods are the most effective modern contraceptive methods for preventing unintended pregnancy. NuvaRing and Ortho Evra patches are not included in LARC as they are used on monthly basis.

Table 10.2: Progestogen only contraceptives.

Route of administration	Contraceptive methods	Progestin
Oral	Minipill	Levonorgestrel, norethisterone, ethynodiol diacetate, Norgestrel, Lynestrenol, Desogestrel
Injectable	Depo-Provera (DMPA)	Medroxyprogesterone acetate
	NET-EN	Norethisterone
Subcutaneous implants	Norplant	Levonorgestrel
	Jadelle	Levonorgestrel
	Implanon	Desogestrel
Intrauterine device	Mirena	Levonorgestrel

Progestogen-only pill (POP)

There had been a constant search for progestins only contraceptive for women in whom the combined pills were unsuitable. The POP containing levonorgestrel had drawbacks like higher failure rate, androgenic side effects and unfavorable metabolic effects. The newer gonane, desogestrel was better than levonorgestrel as it was free of these problems. The POP containing 75 µg desogestrel acts through its active metabolite-etonogestrel. Etonogestrel has a high binding affinity for progesterone receptors and very low affinity for androgenic receptors. It has a high selectivity index. The contraceptive effect is exerted by inhibition of ovulation in up to 98.3% of cases and by cervical mucus changes in up to 95-99% of cases. There is not much suppression of gonadotropins and 40% of women ovulate.

The drug should be taken from the first day of cycle. The peak effect on the cervical mucus are seen after 2-6 hours of taking the drug. The pill should be taken within three hours of the scheduled fixed same time each day. A delay of more than 12 hours should be treated as a missed pill and another method like barrier contraception should be advised for 48 hours. The contraindications include current deep vein thrombosis, severe liver disease, malignancy and hepatic enzymes-inducing drugs. The most common side effects are alteration in the bleeding pattern, acne, headache, nausea, breast pain and vaginitis. The changes in lipid profile and carbohydrate metabolism are minimal with desogestrel as compared to the older gonanes.

Major advantages of POPs are in lactating women and in older women. The failure rate of POP is comparable to COCs and the Pearl index is reported to be 0.41. The ongoing pregnancy rate is 5% because of missed pills. Studies have shown that there is suppression of luteinizing hormone (LH), reduced serum progesterone concentration and decrease in follicular growth from 20-30 mm to 9-20 mm. The ectopic pregnancy rate is between 6% and 10% of unintended pregnancies. Desogestrel and lactation both suppress ovulation, so there is a delay in menstruation and and the contraceptive effect is increased due to synergistic action.

Emergency Contraceptive Pill (ECP)—Levonorgestrel

All the hormonal contraceptive pills can be used as ECPs in varying doses. However, drug controller of India has approved only Levonorgestrel. ECPs can be used within and up to 72 hours of unprotected coitus. Levonorgestrel is available as 0.75 mg tablet. It is given as two doses at 12 hours interval. The mechanism of action is similar to POPs. It is effective, safe, nonteratogenic and does not affect lactation.

The effectiveness decreases with frequent use. Side effects include nausea, vomiting, irregular bleeding, breast tenderness, headache and dizziness. It can be given at any time in the menstrual cycle. If vomiting occurs, then the tablet should be repeated. The failure rate is 2% and in these cases continuation of pregnancy is advisable as it is not teratogenic.

In a recent clinical study of three regimens for emergency contraception within 120 hours of unprotected coitus, a multicentric trial was undertaken in 4,136 women. Women were randomly assigned to one of the three regimens—10 mg single dose mifepristone, 1.5 mg single dose levonorgestrel and two 12-hourly doses of 0.75 mg levonorgestrel. All the three regimes were equally efficacious in preventing unwanted pregnancy. There was no difference in the two regimens of levonorgestrel and it was concluded that single dose regimen of the drug can substitute the two-dose regimen.

Recently Ulipristal Acetate has been approved for emergency contraception and medical management of uterine fibroids. It is a Selective Progesterone Receptor Modulator. It is a 30 mg tablet to be used within 120 hours (5 days) after an unprotected intercourse or contraceptive failure. It has been shown to prevent about 62–85% of expected pregnancies, and prevents more pregnancies than emergency contraception with levonorgestrel. Ulipristal acetate is available by prescription for emergency contraception in over 50 countries.

Long-acting Contraception

The reversible and short-acting methods of contraception have the problem of high failure rates and compliance difficulties. These are as effective as sterilization and IUCDs and more effective than oral and barrier contraceptives because compliance is not dependent on frequent supply or instruction in use. However the requirement of a trained person for usage is there. Two effective methods:
1. Injectable Depot-Provera
2. Contraceptive implants.

Depot-Provera

It has been part of contraception program of many countries since three decades and is found to be safe, acceptable and effective. It is a long-acting injectable method with the action of a sustainable release system like the contraceptive implants. The dose is 150 mg IM every 3 months in buttocks or arm and it is more than 99% efficacious in preventing pregnancy. It is not recommended for those who have unexplained vaginal bleeding, liver disease, breast cancer or stroke. It should be used with caution in teens, and in women with osteoporosis because of its relation to bone loss. Side effects include irregular menstrual periods, amenorrhea (50% women at 1 year), headache, depression, acne, weight gain etc. Fertility/

ovulation is delayed for up to 6–18 months following the injection (details shown in DMPA).

Contraceptive Subdermal Implants (Table 3)

1. Norplant
2. Norplant-2/Jadelle/Sinoplant II
3. Implanon.

Mechanism of Action

1. The progestins suppress the LH surge at both hypothalamic and pituitary level. About one third of all cycles in Norplant users are ovulatory. They cause luteal insufficiency in the cycles which are ovulatory. Folliculogenesis and follicular development occur thus avoiding hypoestrogenic adverse effects. Implanon, however, exerts contraceptive effect mainly by inhibition of ovulation and it inhibits ovulation for at least 2.5 years.
2. The steady levels of progestin decreases the amount and thickens the cervical mucus making it impenetrable and hostile to sperms.
3. They decidualize and atrophy the endometrium, thus inhibiting implantation of the fertilized ovum. However, no evidence of fertilization was observed in Norplant users.

Table 10.3: Contraceptive implants available or being developed.

Implant	Distinctive components	Registration	Lifespan	Failure rate (pregnancies/year)
Norplant	6 silicone capsules releasing LNG	In about 60 countries	7 years*	<1
Jadelle	2 silicone rods releasing LNG	In some European countries, USA, Thailand and Indonesia	5 years	<1
Implanon	1 polymer (resin) rod releasing etonogestrel	Australia, Indonesia and many European countries	3 years	<1
Nestorone	1 silicone rod releasing nestorone	Brazil	2 years	<1
Nexplanon	1 rod releasing etonogestrel		3 years	<1

*Approved for 5 years.

The progestin from the implant diffuses into the surrounding tissues and it is absorbed by the blood vessels and distributed systemically throughout the body via the circulatory system. With Implanon the circulating levels of etonogestrel are high enough within 8 hours of insertion to prevent ovulation. Progestin levels with Norplant and Jadelle are more variable than with Implanon and with them a backup method is required for up to 3 days after insertion.

The circulating progestin levels are affected very little by body weight in Implanon users, whereas this is not so in Norplant and Jadelle users of weight more than 154 lb. Despite that the release rate is high enough to prevent pregnancy as reliably as oral contraceptives. Implanon may be a better choice in obese women. The levels are also affected by SHBG levels with levonorgestrel having a higher affinity than etonogestrel for SHBG.

Some alterations in the menstrual pattern occur in the first year of use in approximately 80% of implant users, later decreasing to about 33% of users by the fifth year. The changes include irregular bleeding, menorrhagia, oligomenorrhea, spotting, polymenorrhea and amenorrhea. Amenorrhea is because the local progestational effect causing endometrial regression and atrophy. The bleeding is due to increased number, tortuosity and size of the blood vessels which are fragile in the face of rapid endometrial regression to an atrophic state. With Implanon amenorrhea is more common (21% of users in first year, 30-40% after one year) than with Norplant (10% after first year). Bleeding is lighter and less frequent among Implanon users because lesser estrogen-induced endometrial proliferation owing to more profound ovarian suppression. Blood hematocrit levels show an increase overall in implant users because of a decrease in the average amount of menstrual loss.

In cases of prolonged bleeding with the implants, estrogens are advisable in a short course: estradiol 2 mg or conjugated estrogens 1.25 mg daily for 7 days.

Norplant

Norplant was developed by the Population Council and manufactured in Finland in 1983. It was approved in the USA in 1990 and marketed in 1991. In 2002 it was withdrawn for business reasons and not medical reasons. It is a sustained-release system consisting of 6 rods or capsules, each containing 36 mg of levonorgestrel in crystal form and measuring 34 mm in length and 2.4 mm in diameter.

The silastic tubing of the capsule is permeable to the progestin within. It was found to be a stable system even after 9 years of use. Its average daily release rate is 85 μg in the first month, 35 μg in the first year and 30 μg in the second year of use.

Norplant-2 or Jadelle

Its Chinese version is called Sinoplant II. It was also developed by Population Council and manufactured in Finland. It was approved in USA in 1996, but never marketed. It is a two-rod system with silastic tubing, each containing 75 mg levonorgestrel and measuring 43 mm in length and 2.5 mm in diameter. Thus the total is 66 mg less levonorgestrel and in the form of a mixture with dimethylsiloxane. The efficacy and side effects are similar to Norplant, but the removal is faster and easier. The release rates are higher than Norplant with 100 μg, 40 μg and 30 μg in the first month, first year and second year respectively.

Implanon

It is a single rod system measuring 40 mm in length and 2.0 mm in diameter and containing 68 mg of etonogestrel or 3-keto desogestrel dispersed in a core of ethylene vinyl acetate wrapped with a 0.6 mm thick membrane of the same material. The initial daily release rates of 67 μg decrease to 30 μg after two years. It provides contraception for at least three years. Side effects are similar to Norplant and Jadelle with a lesser bleeding and higher rate of amenorrhea.

Nexplanon is a hormone-releasing birth control implant for use by women to prevent pregnancy for up to 3 years. The implant is a flexible plastic rod about the size of a matchstick that contains a progestin hormone called etonogestrel. It is 4 cm long by 2 mm wide, containing 68 mg of etonogestrel and also contains a small amount of barium sulfate so that the implant can be seen by X-ray, and may also contain magnesium stearate. A single nexplanon implant for up to 3 years. It is US FDA approved.

Efficacy: In studies with Norplant, the pregnancy rate was 0.2 per HWY. If luteal phase insertions are excluded then the pregnancy rate is as low as 0.01/HWY. The contraceptive efficacy of Implanon surpasses that of sterilization and the Pearl index is 0.01. There is an immediate contraceptive effect when inserted within the first 7 days of the cycle. It is safe, highly reliable, easy to use, rapidly reversible unlike injectables. Women can plan their pregnancy precisely as

Clinical Usage Guidelines of Progestins

return of fertility occurs within a few weeks after removal. It is not associated with effects on carbohydrate and lipid metabolism, kidney and liver function or immunoglobulin levels. There are no weight restrictions for Implants—in slender women no pregnancy was reported till 7th year of use; in heavier women till 5th year pregnancy rates were lower than with OCPs. Implanon can be used in obese women with equal efficacy. In adolescents it is better than oral route with higher continuation rates.

Disadvantages

1. It requires a trained skillful person for insertion and removal. Women are dependent on the availability of a clinician. The short-term costs are higher including the costs for initiation, discontinuation, cost of implants and fees of the surgeon. But the costs are comparable to any other method over a usage period of five years.
2. It causes disruption of bleeding patterns, especially in the first year of use. This is because estrogen levels are nearly normal and in the absence of progestin withdrawal, the endometrium sheds at unpredictable intervals.
3. The implant rods are visible under the skin which may be unacceptable to some couples.
4. Removal can be complicated if the rods are displaced and the incidence of complicated removals is around 5% with Norplant but lower with Jadelle and Implanon.
5. Ectopic pregnancy—about 30% of Norplant pregnancies are ectopic. If pregnancy is suspected with an implant, ectopic pregnancy should be the first diagnosis, though the overall risk of developing a pregnancy is low.

Ideal Candidates for Contraceptive Implants

- Women who want to delay their next pregnancy for at least 2-3 years
- Lactating women who intend to breastfeed for 1-2 years
- Women with menorrhagia and anemia
- Women who have completed their family but are indecisive about permanent sterilization
- Women who experience side effects with estrogen within combined contraceptives
- Women who are not compliant with daily contraceptive usage

- Women with contraindications for estrogen contraceptives
- Women who have contraindications for IUDs
- Women who want coitus-free method of contraception
- Women with chronic illnesses
- Women desirous of long-term and effective contraceptive choice
- In couples in whom elective abortion is unacceptable.

Contraindications

- Absolute:
  - Active thrombophlebitis or thromboembolic disease
  - Acute liver disease
  - Benign or malignant liver tumors
  - Suspected or known breast cancer
  - Undiagnosed genital bleeding
- Relative:
  - History of ectopic pregnancy
  - Heavy smoking (>15/day) in women above 35
  - Hypercholesterolemia
  - History of cardiovascular disease, myocardial infarction, coronary artery disease, angina, previous thromboembolic event, cerebrovascular event, artificial heart valves.
  - Hypertension
  - Gallbladder disease
  - Diabetes mellitus
  - Chronic disease, immunocompromised patients
- Implants are not recommended in:
  - Women with severe acne
  - Severe migraine or vascular headaches
  - Severe depression
  - Women using microsomal liver enzymes inducing medicines, e.g. carbamazepine, felbamate, rifampicin, phenobarbital, phenytoin, nevirapine, etc.

Hormonal IUCDs

Mirena (LNG-IUS)

This is an intrauterine progestin device releasing levonorgestrel. The hormone reservoir is around the vertical polyethylene stem and consists of a mixture of 52 mg of levonorgestrel and silicone (polymethylsiloxane). The hormone is released into the uterine cavity at a rate of approximately 20 µg per day. A stable plasma level of 150-200 pg/mL is achieved within a few weeks of insertion. This level is lower than that achieved by administration of systemic progestin. The contraceptive effect of LNG-IUS is by inducing local progestogenic changes in the endometrium, which include stromal pseudo-decidualization, glandular atrophy, leucocyte infiltration and stromal mitosis along with the sperm-hostile effect on cervical mucus. The vertical stem of this system is thicker so there is increased insertion pain and risk of perforation. Due to the local effects of the LNG, correct fundal placement is not crucial.

The contraceptive efficacy has been assessed in a randomized comparative trial over five years. The intrauterine and extra-uterine pregnancy rates were lower than that for Nova-T. The Pearl index after five years was 0.09/HWY and ectopic pregnancy rate was 0.02/HWY. The continuation rate was higher at 66% at the end of second year of use.

With this system, there is progressive reduction in menstrual blood loss and duration. Many women have amenorrhea with Cameron reporting that 20% women became amenorrhoeic after 6 months of use and 50% after 5 years of use. Apart from this benefit, there are many other health benefits associated with this system, e.g. a decrease in the incidence of anemia, PID, endometrial hyperplasia and reduction in fibroids, endometriosis and adenomyosis. Published the results of use of LNG-IUS system in the treatment of endometriosis and adenomyosis. Dysmenorrhea, dyspareunia, menstrual flow and uterine volume were decreased with a significant increase in the hematocrit in their patients. It is especially useful in perimenopausal women and in women with idiopathic menorrhagia. For the treatment of menorrhagia, this system has been evaluated the world over with favorable results. The mean blood loss of 176 mL was reduced to 24 mL after three months of use, to 15 mL after six months of use and to 5 mL after 12 months of use. The superiority of Mirena over norethisterone, tranexamic acid and flurbiprofen has been documented. Regarding its comparison to TCRE, there was 98% reduction in bleeding one year after TCRE and 90% reduction after one year of Mirena. The present uses of this system include treatment of abnormal uterine bleeding (AUB), fibroids, adenomyosis and endometriosis. This system has been recommended as progestin supplementation in women on hormone replacement therapy (HRT). It has also been studied for the treatment of premenstrual syndrome (PMS) and dysmenorrhea.

Table 10.4: Various hormone replacement therapy regimes.

Regimens	Estrogens	Progestins
Cyclic Sequential	D 1-25	D 13-25
Continuous sequential	Daily	D 1-14
Continuous combined	Daily	Daily
Cyclic combined	D 1-25	D 1-25

Side effects include irregular bleeding, breakthrough bleeding specially in the first six months and other side effects of progestins.

Hormone Replacement Therapy

In the perimenopausal and postmenopausal years, women suffer from various symptoms due to estrogen insufficiency. Therefore estrogen replacement therapy becomes necessary to get relief from these menopausal symptoms. But unopposed estrogen increases the risk of endometrial hyperplasia and malignancy and breast carcinoma. For this reason, addition of progestogens becomes essential to prevent these effects. Progestins also suppress hot flushes, cause withdrawal bleeding and prevent postmenopausal blood loss. The various HRT regimes are given in Table 10.4.

Role of Progestins in HRT

1. *Osteoporosis*: Progestins alone help to prevent osteoporosis, but when added to estrogens, they promote new bone formation and restore the lost bone mass.
2. *Prevention of endometrial cancer*: Unopposed estrogen therapy increases the risk of endometrial cancer by 5-10 fold. Addition of progestins not only negates this risk, but also decreases this risk to less than that in untreated women. In the Wilford Hall United States Air Force Medical Center studies, the incidence of endometrial cancer was 390.6:100,000 women in unopposed estrogen users; 49:100,000 in estrogen-progestin users and 245.5:100,000 in the untreated women. Annual endometrial biopsies need not be performed in women having regular withdrawal bleeding from estrogen-progestin therapy. Progestin Challenge Test is the most useful test for determining women at an increased risk of endometrial cancer.
3. *Treatment of endometrial hyperplasia*: Unopposed estrogens cause incomplete shedding of the endometrium and thus have a

role in development of endometrial hyperplasia and cancer. The incidence of adenocarcinoma of the uterus rose to 30% by the tenth year of estrogen therapy in Gusberg's study. It is essential that patients with endometrial hyperplasia have a repeat curettage after six months of progestin therapy. Norethindrone acetate in the dose of 5 mg seems to be more effective than medroxyprogesterone acetate 10 mg. The minimum dosage and duration of C-21 progestin that will protect endometrium from hyperplasia is 10 mg for 12-14 days.
4. *Reduction of breast cancer risk*: The role of HRT in reducing the risk of breast carcinoma is controversial. In hysterectomized women, progestins are not added to estrogen replacement therapy. A few studies, however show that adding progestins increases this risk depending on receptor positive status in breast cancer tissue.
5. *In hysterectomized women*: Progestins are administered in these women for the following indications:
   - Past history of endometriosis
   - Previously treated for endometrioid tumors of ovary
   - Procedures that leave residual endometrium behind, e.g. supracervical hysterectomy, endometrial ablation
   - High risk of osteoporosis with increased triglycerides.

Newer Additions in HRT

Combination of 17β-estradiol and dydrogesterone has been launched with the name of Femoston (Femoston-conti and Femoston). It contains 17β-estradiol in the dose of 1-2 mg and dydrogesterone 5-10 mgs. It is metabolically friendly, having no effect on lipid and carbohydrate metabolism. It is also not androgenic.

Estrogens : 0.625 mg Conjugated estrogen
2 mg Estradiol valerate
1 mg Micronized estradiol
50 μg Transdermal estradiol

Progestins
(Two weeks : 200 mg Micronized progesterone
every month) 10 mg MPA
10 mg Dydrogesterone
(Daily) : 0.35 to 1 mg Norethindrone
100 mg Micronized progesterone
2.5 mg MPA

Adverse Effects of Adding Progestins to ERT

Effects on lipid profile: Presumed adverse effects on lipids and lipoproteins may be seen thus negating the most important benefit of ERT viz. prevention of cardiovascular disease. In the 3-year Postmenopausal Estrogen Progestin Intervention Trial, estrogen alone or in combination with a progestin improved lipoprotein levels and lowered fibrinogen levels. There are no long-term effects of added progestins, especially when adequate doses of estrogen are given.

Other effects: Progestins mainly cause premenstrual syndrome like symptoms such as headache, abdominal bloating, irritability, depression, lethargy, edema, and breast tenderness. However, with the use of mild diuretics and change in doses, types of progestins or route of administration almost all women can use progestins without any adverse effects.

Abnormal Uterine Bleeding

Progestogens help in abnormal uterine bleeding (AUB) by arresting acute bleeding, making the menstrual cycle regular and protecting the endometrium from hyperplastic effects of estrogen. It acts by the following complex mechanisms:
- It reduces the endometrial vascularity
- It reduces fibrinolysis
- It prevents glandular growth by decreasing estrogen metabolism
- It stimulates lysosomal enzymes release. Phospholipase helps in formation of thromboxane A2 leading to platelets aggregation
- It promotes local stromal support
- It inhibits pituitary-ovarian function, thus preventing endogenous ovarian steroid production.

In most cases of anovulatory bleeding, progestin therapy will control the bleeding once uterine pathology is excluded. Failed progestin therapy is strongly suggestive of other pathology causing the problem and calls for additional diagnostic evaluation. In oligomenorrheic anovulatory women with episodic abnormal bleeding, cyclic progestin treatment with an orally active progestin like medroxyprogesterone acetate 5–10 mg daily for 12–14 days each month will induce an orderly regular predictable self-limited progesterone withdrawal bleeding. Progestins can be given beginning first day of every month or beginning 15 days after the onset of the last progestin induced menses. This therapy may not sufficiently suppress the hypothalamic-pituitary axis and prevent random ovulation.

Treatment with cyclic estrogen-progestin contraceptive is better for women who ovulate infrequently and are sexually active. The volume and duration of bleeding and the associated dysmenorrhea decreases progressively. Longer duration of treatment results in lighter and fewer menses with the only disadvantage of episodic breakthrough bleeding.

Acute severe bleeding can be managed by high-dose progestin alone given for three weeks, starting with medroxyprogesterone acetate 10-20 mg twice daily or megestrol acetate 20-40 mg twice daily or norethindrone 5 mg twice daily, provided that the endometrium is normal or increased in thickness, and then reduced to once daily treatment after 7-10 days. Upon progestin withdrawal, a substantial amount of endometrium remains to be shed causing continued bleeding or "medical curettage". Therefore patients need to be forewarned regarding this. Following this, standard cyclic progestogen alone or an estrogen-progestin combination are prescribed for the long term. Depo-medroxyprogesterone acetate is advisable as maintenance therapy for women in whom combined pill is contraindicated or not feasible.

19-norethisterone derivatives like norethisterone acetate 5-10 mg thrice daily is given for 5-7 days in cases of acute menorrhagia. Subsequently progesterone or its derivatives like dydrogesterone or MPA is given from day 14 to 26 following the withdrawal bleeding. This is continued for the next few cycles.

Combined estrogen and progestin therapy is another option for AUB. If contraception is also desired, oral contraceptive is a better choice. For acute bleeding low dose OCP is given twice daily for 5-7 days and then continued once a day for 3-6 cycles. If the patient does not want contraception or is not in the reproductive age group, then estrogen therapy is followed by progestin therapy from day 14 to 26.

If bleeding is prolonged with progestin therapy alone, then estrogen breakthrough bleeding is the likely cause. In such cases, tissue biopsy yields minimal tissue. The treatment for these cases is conjugated estrogens 1.25 mg or estradiol 2 mg daily for 7-10 days, followed by estrogen plus MPA 10 mg daily for the next 10 days. If the acute bleeding is moderately heavy, estrogen is administered every four hours during the first 24 hours. For heavy bleeding, high doses of conjugated estrogens 25 mg IV 4-hourly is given until bleeding decreases, which is then followed by oral estrogen-progestin combination therapy. If there is no response in 24 hours, curettage is suggested. DMPA and Mirena has also been used in DUB.

In the perimenopausal woman with abnormal bleeding, endometrial biopsy or hysteroscopy is essential to rule out organic disease. If findings suggest proliferative or hyperplastic endometrium uncomplicated by atypia or dysplasia, then cyclic oral progestin therapy must be administered–micronized progesterone 200 mg or medroxyprogesterone acetate 5-10 mg daily for D5-25 of each month. If hyperplasia is present then after 3-4 months of similar therapy, endometrial aspiration curettage is essential. Monthly cyclic progestational therapy reverse simple hyperplastic changes in 95-98% of women and is able to control irregular bleeding. Therapy should be continued until withdrawal bleeding ceases, i.e. until the onset of menopause. Then estrogen can be added as HRT if there is no histological regression, alternate therapeutic surgery should be contemplated. It must be remembered that progestins can mask abnormal tissue and therefore follow-up biopsy must be done after 3 months of progestin therapy. Despite progestational treatment, progression or persistence of abnormal endometrium is seen in 28.4% women with complex hyperplasia and in 26.9% women with atypical hyperplasia. A minimum of 3-6 months of progestational therapy must be tried before deciding for hysterectomy.

Endometrial Cancer

Progestins are most commonly used adjuvant or palliative therapy in early, advanced or recurrent endometrial cancer. They can revert or counteract the stimulatory effect of estrogen (antimitotic) by:
a. Increasing the clearance of estrogen from body by enzyme induction.
b. Decreasing estrogen mediated transcription of oncogenes.
c. Decreasing estrogen induction of its own receptors.

MPA and megestrol acetate are the most commonly used agents. These drugs are continued for at least 3 months and remain an important alternative for women who are unable to tolerate chemotherapy.

Regimes

- Medroxyprogesterone acetate 1 gm/week or 150 mg/day orally
- Megestrol acetate 160 mg/day orally
- 17 Hydroxyprogesterone caproate 1 gm/week IM.

Luteal Phase Defect (LPD)

Luteal phase defect is an ovulatory disorder of considerable clinical importance that is implicated in 3-20% cases of infertility and 5-60% cases of recurrent abortions. As a subtle disruption of ovulatory or luteal function, it may be the most common ovulatory disorder in women. Pathophysiologic alterations of the complex reproductive process lead to delayed endometrial maturation. The characteristics of LPD include disordered folliculogenesis, defective corpus luteum function and abnormal luteal rescue by the early pregnancy. In the absence of an identifiable correctable underlying cause of LPD, progesterone replacement and clomiphene citrate are the usual treatment options for consideration. Combination therapy, gonadotropins and other treatments are reserved for refractory cases. Luteal phase support is essential in IVF cycles specially those involving GnRH agonist down regulation.

The causes of luteal phase insufficiency in assisted reproductive technology (ART) cycles with super ovulation by gonadotropin-releasing hormone (GnRH) are supraphysiological levels of estradiol in follicular phase leading to luteolysis, ineffective luteal cells due to puncture of follicle and insufficient LH due to ineffective corpus luteum by pituitary desensitization.

Natural progesterone is the drug of choice for this condition although 17α-hydroxyprogesterone is also used in patients with recurrent abortions. Other synthetic progestins are limited by their potential teratogenic effects, adverse effects on lipid profile and psychological side effects.

The preferred route of administration is per vaginal, which induces secretory transformation of the endometrium at the lowest dose by virtue of uterine first pass effect plus having less systemic side effects. Pregnancy rates of 35-70% in infertility patients and of 80-85% in patients with recurrent pregnancy loss have been reported by the use of vaginal progesterone for the treatment of LPD.

Vaginal progesterone is available as 8% gel Crinone, which delivers 90-180 mg daily. It is also available in the form of pessaries containing micronized drug in polycarbophil base mimicking negatively charged mucin which attaches to the vaginal epithelium. This increases its absorption and bioavailability plus also ensures its release over a prolonged duration of period owing to the vaginal reservoir effect.

Intramuscular progesterone achieves the highest plasma levels, adequate secretory transformation and satisfactory pregnancy rates of 15-60%. It is limited by the requirement of prolonged daily injections and local inflammation with occasional abscess formation at the injection sites.

Oral progesterone, although most convenient, has been found to be ineffective in inducing secretory phase endometrium, has systemic side effects like sedation, drowsiness, flushes and nausea and has poor pregnancy rates of just 5-30% only. The consensus regarding micronized progesterone given as a vaginal gel in the dose of 180 mg or as a 300 mg pessary daily for LPD has been refuted over the recent years.

Dydrogesterone is used in dosage of 30 mg/day with satisfactory results. Administration of hCG in the dose range of 500-2000 IU IM twice a week after ovulation is an indirect method of luteal support. The treatment for luteal phase support is usually for a period of 4-14 weeks depending on the nature of diagnosis and indication for the treatment.

Premenstrual Syndrome

Premenstrual symptoms sufficient to impair daily life affect up to 40% of women in the reproductive age group, with severe impairment observed in 3% of these women. It results as a direct consequence of the hormonal events of the normal ovarian cycle in women. There is no conclusive evidence to suggest that there is any difference in the ovarian hormones status in PMS patients as compared to that in non-suffering women.

Various medical therapies have been advocated for PMS with varying results. Based on the estrogen dominance or progesterone deficiency theory, progestogens have been tried. In the past progesterone was used commonly for PMS symptom relief in the form of suppository or injection. Micronized progesterone, 200 mg and dydrogesterone, 10 mg, daily for a period of 12 days have been studied over the years with satisfactory results (confirm results). The added advantages of progestins in PMS are many. These include endocrinal support in the elderly women, endometrial protection from the risk of cancer, cardiovascular and skeletal benefits. However, large double-blind placebo-controlled trials, a meta-analysis and a recent Cochrane systematic review concluded that the effects of progesterone are similar to those of placebo and that it is not an effective treatment for PMS.

Postponement of Menstruation

Due to societal and personal reasons, especially in conservative societies, women may need some manipulation of their cycles. For this purpose, medroxyprogesterone acetate 10 mg daily or norethisterone 5 mg thrice a day can be administered at least 3 days prior to the expected date of menstruation and continued till desired. Following withdrawal of the progestins, menses start within 48–72 hours.

Endometriosis

Continuous use of progestogens causes decidualization of the endometrium, whether in-situ or in ectopic/extra-uterine sites, resulting in atrophy of the endometriosis deposits. Dydrogesterone or MPA are administered in high doses, i.e. 20–30 mg/day to achieve amenorrhea and are continued for 6–9 months. Intramuscular DMPA can be given, but is usually avoided in infertile women due to concerns regarding delayed fertility post its use. Progestins have been found to have similar efficacy to that of danazol and are also cheaper than danazol.

Out of all medical therapies dienogest has proved to be most promising. The dose is 2 mg. The potency is 5 times to that of levonorgestrel and 10 times to that of medroxyprogesterone acetate. It is a nortestosterone that has cyanomethyl group instead of an ethinyl group at 17 position. It has both properties of nortestosterone and derivatives of progesterone. Treatment by dienogest alleviates the pain, decreases endometriotic lesion and improves the quality of life. It is safe, tolerable and does not decreases patient's compliance. It does not have androgenic or mineralocorticoid activity. There is no effect on lipid or carbohydrate metabolism. The action on endometriotic lesion is due to hypoestrogenic and hypergestogenic effect causing decidualization and atrophy. Fertility returns averagely in 1–43 days.

Preterm Labor and Preterm Birth

Progesterone blocks the oxytocic effect of PGF2α and α-adrenergic stimulation on the myometrium, therefore it increases the tocolytic response in the pregnant myometrium. Hence, if used in the prevention of preterm labor, it may improve perinatal outcomes by reduction in perinatal morbidity and mortality.

In a placebo-controlled study in 77 pregnant women, Johnson et al. demonstrated that 17α-hydroxyprogesterone caproate is effective in the prevention of premature delivery in women at risk and also resulted in increased birth weights. In this study 250 mg of the drug was administered weekly by the intramuscular route from 28th gestational week onwards.

In another study in 57 women, oral micronized progesterone in the dose of 400 mg daily was administered daily from the 30th week of pregnancy onwards and the uterine activity measured. In the progesterone group, 73% women experienced a decreased frequency of myometrial contractions as compared to 50% women in the placebo group. The plasma progesterone levels increased from a mean level of 101.6 ng/mL to a mean of 152.19 ng/mL in the treatment group. The oral drug was well tolerated and there were no major side effects, except minimal drowsiness.

Progesterone suppositories containing 100 mg of progesterone were used in some studies in preterm labor. The pessaries were used daily night from 20th to 34th week of gestation. The incidence of preterm delivery in the pessary group was 13.8% as compared to 28.5% in the placebo group.

Natural progesterone is the drug of choice in preterm labor as it does not have any teratogenic, hemodynamic or metabolic adverse effects on the pregnancy or the fetus, which are oftentimes seen with the use of betamimetic and synthetic drugs. In current clinical practice, micronized progesterone or dydrogesterone can be used in those cases of preterm labor where there is no response to standard therapy or there are worrisome side effects with use of tocolytics.

Allylestrenol or 17α-allylester-4-en-17β-ol is a 19-nortestosterone derivative and is among the estrange category of progestins. It has been in use for more than 40 years. It has no androgenic, endocrinological, estrogenic or anabolic side effects due to the absence of side chain at C-17 and it has an angular methyl group at C-19 which makes it orally effective. It is a known placentotrophic agent causing an increase in many placental enzymes due to stimulation of β2-adrenergic receptors in the placenta. It stimulates certain syncytiotrophoblast enzymes, which include DPNH-diaphorase, cysteine aminopeptidase and delta-5, 3β-hydroxysteroid dehydrogenase. Following its administration, there is an increase in the secretion of hCG, hPL, estriol and progesterone. It causes myometrial relaxation, increased placental blood flow and better fetal metabolic exchange. The therapeutic dose is 15–30 mg/day. After its intra-amniotic administration of 50–200 mg, there is a definite rise in plasma progesterone (+80%) and a moderate rise in hPL (+20%) with increased secretion of urinary pregnanediol and total estrogen.

Vaginally dosed progesterone is being investigated as potentially beneficial in preventing preterm birth in women at risk for preterm birth. The initial study by Fonseca suggested that vaginal progesterone could prevent preterm birth in women with a history of preterm birth.[61] According to a recent study, women with a short cervix that received hormonal treatment with a progesterone gel had their risk of prematurely giving birth reduced. The hormone treatment was administered vaginally every day during the second half of a pregnancy.[62] A subsequent and larger study showed that vaginal progesterone was no better than placebo in preventing recurrent preterm birth in women with a history of a previous preterm birth, but a planned secondary analysis of the data in this trial showed that women with a short cervix at baseline in the trial had benefit in two ways: a reduction in births less than 32 weeks and a reduction in both the frequency and the time their babies were in intensive care.[63,64] In another trial, vaginal progesterone was shown to be better than placebo in reducing preterm birth prior to 34 weeks in women with an extremely short cervix at baseline.[65] An editorial by Roberto Romero discusses the role of sonographic cervical length in identifying patients who may benefit from progesterone treatment.[43] A meta-analysis published in 2011 found that vaginal progesterone cut the risk of premature births by 42% in women with short cervixes.[44] The meta-analysis, which pooled published results of five large clinical trials, also found that the treatment cut the rate of breathing problems and reduced the need for placing a baby on a ventilator.[45]

Puerperal Depression

Postnatal depression affects 10–15% of women following childbirth and persists for more than one year in 40% of these patients.

A study was conducted in 2000 to find out the effect of gonadal steroids in women with history of puerperal depression. It is well known that during pregnancy there is a supraphysiologic rise in the levels of gonadal steroids which are suddenly withdrawn after delivery. Sixteen euthymic women, 8 with a history of postnatal depression and 8 without any history of postnatal depression, received leuprolide acetate, estradiol and progesterone for 8 weeks and then the steroids were withdrawn. Five of 8 women with a history developed significant mood symptoms during withdrawal period, while no one in the comparison group developed any withdrawal symptoms. This data provides direct evidence in support of the involvement of withdrawal of reproductive hormones estrogen and progesterone in the development of postpartum depression.

In another study it was found that changes in some women depressive mood changes were caused by certain changes in the hormonal axis. Wagner has postulated that the high circulating levels of estrogen during pregnancy protect against depression and that the rapid decline of levels remove this protective effect. A number of studies used HRT in postmenopausal patients have reported an improvement in depression, which gives an indirect evidence. There are also data to support the hypothesis that therapy with 200 µg transdermal estradiol significantly reduces depression scores and accelerates recovery in women with postnatal depression. Natural progesterone use has also been advocated in the dose range of 200–400 mg/day for 7 days postpartum, alone or in combination with estrogen. Uncontrolled studies report benefit of progesterone therapy. A review of literature in 2001 reported that there is not much role of progestins in the treatment of recurrences of postpartum depression. More randomized controlled studies are recommended to prove the efficacy of progestins in this condition.

Threatened and Recurrent Abortions

Progestogens have been extensively used in the past to prevent threatened and recurrent abortions because of their ability to diminish myometrial contractions. However, there is no evidence to support this role. A potential risk of virilization of the female fetus and genital deformities have been reported with such use. The only indication can be a documented serum progesterone deficiency, which is rarely seen in clinical practice.

Other Uses of Progesterone

Assisted Reproductive Technology

Used to support luteal phase, embryo implantation and early pregnancy as part of ART treatment of infertile women. It can be given as:
- Progesterone in oil 50 mg IM daily
- Vaginal suppository 200 mg twice daily
- Micronized progesterone orally 200 mg twice daily usually continued for 14 days.
- *Progesterone 8% vaginal gel*: For women who require progesterone supplementation, administer contents of one prefilled applicator once daily. If pregnancy occurs, continue until placenta autonomy is achieved, up to 10–12 weeks.
- *Progesterone vaginal insert*: One insert (100 mg) 2–3 times daily. Start at oocyte retrieval; continue for up to 10 weeks.

Catamenial Epilepsy

Progesterone can be used to treat catamenial epilepsy by supplementation during certain periods of the menstrual cycle.[46]

Multiple Sclerosis

Progesterone is being investigated as potentially beneficial in treating multiple sclerosis, since the characteristic deterioration of nerve myelin insulation halts during pregnancy, when progesterone levels are raised; deterioration commences again when the levels drop.

Medical Abortifacient

Antiprogestins and selective progesterone receptor modulators (SPRMs), such as mifepristone, can be used to prevent conception or induce medical abortions.

Hidradenitis Suppurativa

Progesterone is starting to be used in the treatment of the skin condition hidradenitis suppurativa.

Trans-Women

Progesterone is sometimes employed as a component of hormone replacement therapy for trans-women.[66]

CHAPTER 11

Role of Progestins in Combined Hormonal Contraception

Hormonal contraceptives are female sex steroids, synthetic estrogens and synthetic progestins.

These can be administered in the form of oral contraceptive pills (OCPs), patches, injectables and vaginal rings. The most commonly used hormonal contraceptive is the combination OCPs containing estrogens and progestins. Combined OCPs can be monophasic and multiphasic. The steroids are given in 21-day or 24-day per cycle. Combination OCPs are packed with 21 active tablets and seven placebos.

The inclusion of placebos allows the user to take 1 pill every day without having to count. In this, pill free period allows withdrawal bleeding to occur and then next pack is started without a break. The 7-day free medication interval was standard for years but studies showed that a shorter medication-free interval is adequate to trigger cyclical withdrawal bleeding and maintains better suppression of ovulation. Ovarian follicles mature more during 7-day medication free interval than during the 4-day interval. Hence, new 24/4 combination could be more effective than 21/7 regime. Other variations of OC administration are the extended cycle and continuous cycle method.

MECHANISM OF ACTION OF ORAL CONTRACEPTIVE PILLS

It prevents pregnancy by following methods:
1. Act on hypothalamic pituitary ovarian axis to suppress synthesis and secretion of folicular stimulating hormone (FSH) and the mid cycle surge of luteinizing hormone (LH) thus inhibiting of development of ovarian follicles and ovulation.
2. Potentiate cervical mucus thickness and viscosity to prevent penetration of sperms.
3. Inhibit implantation of blastocyst.

Role of Progestins in Combined Hormonal Contraception

- *Monophasic pill*: It contains a constant amount of estrogen and progestin and is called monophasic oral pill. Some popular trade names are Femilon, Novelon, Mala-D, Ovral, etc. (Table 11.1).
- *Biphasic pill*: Biphasic oral contraceptive pill delivers the same amount of estrogen each day while progestin dose is increased during cycle. Different formulations are seen in Table 11.2.
- *Triphasic pill*: Triphasic oral contraceptive pills have three different doses of progestins and estrogen that change approximately every seven days. Formulations are seen in Table 11.3.
- *Quadriphasic pill*: Four phasic oral contraceptive pill provides four different doses of progestins and estrogen during a 28-day cycle. Formulation in market are those comprising of dienogest and estradiol valerate with two tablets of 0 mg/3 mg followed by 5 tablets of 2 mg/2 mg, 17 tablets of 3 mg/2 mg, 2 tablets of 0 mg/1 mg and 2 placebo. Trade name is Qlaira. Management of missed pills is very difficult with quadriphasic pills.
- *Ninety-one-day oral pill*: It provides constant source of estrogen and progestin for 84 days, e.g. seasonale. A typical 91-day pill has 20/30 μg of ethinyl estrabiol and 91/150 Levonorgestrel in each active pill. It reduces the frequency of menstrual period from 13/year to 4/year by changing the regimen of active pill. The package may have placebo pill or low dose estrogen pill for withdrawal bleeding.
- *One yearly pill*: The oral pills containing 20 μg of ethinyl estradiol and 90 μg of Levonorgestrel taken throughout the year does not result in any withdrawal bleeding. The trade name is Lybrel. The drug was launched by 2007. After stoppage of pill the withdrawal bleeding occurs after 12 months.
- *Efficacy of combined oral contraceptive pill (COCP)*: It is 3 pregnancies/1000 women/year with perfect use and it is 9/1000 women/year with typical use.

Table 11.1: Preparation of OCPs (Monophasic).		
Estrogen	Progestin	Trade name
30 μg EE	150 μg Leronorgestrel	Mala-D
30 μg EE	3 mg Drospiranone	Yasmin
20 μg EE	150 μg Desogestrel	Femilon

Table 11.2: Biphasic OCPs.

	Ethinyl estradiol	Norethindrone	Trade name
10 tablets	35 μg	500 μg	
11 tablets	35 μg	1000 μg	Orthonovam
7 tablets	35 μg	500 μg	B-Novam
14 tablets	35 μg	1000 μg	

Table 11.3: Triphasic OCPs.

	Ethinyl estradiol	Norgestimate	Trade name
7 tablets	25 μg	180 μg	Ortho tricyclen
7 tablets	25 μg	215 μg	
7 tablets	25 μg	250 μg	
		Desogestel	
7 tablets	25 μg	100 μg	Cycless
7 tablets	25 μg	125 μg	
7 tablets	25 μg	150 μg	
7 tablets	35 μg	50 μg	Trimerci
7 tablets	30 μg	100 μg	
7 tablets	30 μg	150 μg	
		Gestodene	
7 tablets	30 μg	50 μg	Tridene
7 tablets	40 μg	70 μg	
7 tablets	30 μg	100 μg	

COMBINED HORMONAL ORAL METHODS

- *Estrogen content*: Low dose pill contains 30-35 μg of ethinyl estradiol. Further low strength contains 15-20 μg of ethinyl estradiol. The newer pill Qlaira and Zoely contains 17 B estradiol. These have lower side effects.
- *Types of progestogens*: Older pills having Levonorgestrel and Norethisterone had androgenic side effects. However, these have lower thrombotic risks. Higher thrombotic risks are with Desogestrel, Gestodene and Drosperinone. Zoely contains a new progestin Nomegesterol. Cyproterone acetate used in dianette for acne has intermediate risk for venous thrombosis.

- *Advantages of COCP*:
  - Noninvasive method
  - Menstrual regulator
  - Correction of premenstrual syndrome by Drosperinone
  - Reduction of endometrial, ovarian and colorectal cancer
  - Treatment of certain benign diseases like benign breast diseases, functional cysts of ovary, pelvic inflammatory disease (PID), endometriosis.
- *Disadvantages of COCP*:
  - User dependent
  - Nausea, breast tenderness, breakthrough bleeding, premenstrual syndrome (PMS) like symptoms
  - Does not protect against sexually transmitted diseases (STDs)
  - Late restoration of fertility
  - Venous thromboembolism 2-8/100,000 women
  - Myocardial infarction and stroke
  - Breast cancer and cervical cancer.

COMBINED HORMONAL NON-ORAL METHODS

Transdermal Patch

Mode of action is similar to COCPs. This is a combined estrogen progestogen patch approved by FDA in 2002. It is used in USA by the name of ORTHO-EVRA. It is a 5 cm × 4 cm patch, 20 sq cm in area containing 600-750 µg of ethinyl estradiol and 6000 µg of Norelgestromin an active metabolite of third generation Norgestimate. It releases 20 µg of ethinyl estradiol and 120 µg of Norelgestromin every day. It has to be changed every 7 days. In a month 3 patches are applied and one week is free of patch for withdrawal bleeding. Serum levels are one and half times higher than COCPs. The patch has to be applied on dry, clean and intact skin on upper buttock or abdomen but not on the breast. The patch has to be started on 1st day of menstruation.

Side effects include progestin effects like bloating, nausea, breakthrough bleeding and weight gain. There is little extra risk of venous thrombosis. Skin irritation is major disadvantage. It should not be flushed in the toilet without wrapping. Continuation rates are similar to COCPs. It has the advantage of protection for extra 2 days.

Combined Contraceptive Vaginal Rings

The combined estrogen progestin vaginal ring (CVR) was approved by FDA in 2001. It is transparent flexible ring of 54 mm diameter, made up of ethylene vinyl acetate co-polymer. The ring releases 15 µg of ethinyl estradiol and 120 mg of Etonogestrel: a biologically active metabolite of a third generation progestin Desogestrel per day. It is inserted only once in a month every 3 weekly and one week is kept drug free for withdrawal bleeding. The serum levels of estrogen are lowest as compared to patch or COCPs.

The ring is placed in the vagina from 2-5 days of menstruation. Any backup method is used for 1st cycle. If the ring is out it should be reinserted within 3 hours. The single ring is placed continuously in vagina for 3 weeks and one week is off the ring for menstruation. Side effects are expulsion headache, vaginal discharge and vaginitis. It has good cycle control.

Combined Injectable Contraceptives

It contains both estrogen and progestogens and is not FDA approved. It has been in use since 2003. But there are some issues of manufacturing. It contains natural estradiol instead of synthetic ethinyl estradiol. Two preparation are available: one containing estradiol cypionate 5 mg and medroxyprogesterone acetate (MPA) 25 mg whereas other contains estradiol valerate 5 mg and norethisterone enanthate 50 mg. The preparation has to be once a month that is every four week.

The natural estrogen is less potent and has a shorter half-life compared to synthetic one; so have fewer complications. Moreover, the injectables bypass the first pass through liver hence less liver and gallbladder related side effects are observed. The injectables have to be started from first day of period or within 5 days of period with a backup method of 7 days. It is given in deltoid muscle or gluteus muscle. The effect is carried over for 3-5 days extra than 4 weeks. Side effects are similar to COCPs. The effects on blood pressure lipid and carbohydrate metabolism are not there. Injectables have better compliance, hence as it given once a month. Unlike DMPA it results in regular cycle control and early return of fertility.

Chapter 12

Side Effects, Contraindications and Precautions

SIDE EFFECTS OF PROGESTERONE

Genitourinary

Genitourinary side effects have included breast tenderness (27%), breast pain (6%), breast carcinoma (2%), breast enlargement, urinary problems (11%), vaginal discharge (10%), vaginal dryness (6%), leucorrhea, uterine fibroid, vaginal dryness, fungal vaginitis, vaginitis, dysuria, cystitis, and urinary tract infection, cervical erosion, cervical secretions and mucus changes.

Nervous System

Nervous system side effects have included headache (31%), dizziness (15%), confusion, depression (19%), fatigue (8%), irritability (8%), worry (8%), insomnia, somnolence, cerebral edema, asthenia, pyrexia, increased sweating, nervousness, migraine, tremor, and speech disorder. Syncope (with and without hypotension) has been reported.

Cardiovascular

Chest pain (7%), hypertension, angina pectoris, syncope, and palpitations, cerebral thrombosis. A statistically significant association has been demonstrated between use of estrogen-progestin combination drugs and pulmonary embolism and cerebral thrombosis and embolism. For this reason patients on progestin therapy should be carefully observed.

Gastrointestinal

Abdominal pain (20%), bloating (8%), diarrhea (8%), nausea (8%), constipation (3%), dyspepsia, dry mouth, gastroenteritis, hemorrhagic rectum, hiatus hernia, and vomiting have been reported. The side effects reported during clinical studies evaluating the use of progesterone gel (8%) are constipation (27%), nausea (22%), and diarrhea (8%).

Musculoskeletal

Joint pain (20%), musculoskeletal pain (12%), back pain (8%), arthritis, leg cramps, hypertonia, muscle disorder, and myalgia.

Psychiatric

Depression (19%), anxiety, impaired concentration, insomnia, forgetfulness, and personality disorder.

Respiratory

Cough (8%), bronchitis, nasal congestion, pharyngitis, pneumonitis, sinusitis and respiratory embolism.

Dermatologic

Acne, hirsutism, loss of scalp hair, erythema multiforme, erythema nodosum, hemorrhagic eruption, itching, urticaria, pruritus, rash, skin discoloration, seborrhea, verruca, alopecia and wound debridement during therapy. Additional dermatologic side effects have included case reports of familial autoimmune dermatitis.

Ocular

Ocular neuritis, retinal thrombosis, abnormal vision, proptosis, migraine, papilledema.

Hepatic

Reversible cases of hepatitis, cholestatic jaundice, elevated transaminases, and cholecystectomy have been reported.

Endocrine and Metabolic

Glucose intolerance, galactorrhea, weight gain, amenorrhea, breakthrough bleeding, menstrual flow changes, irregular spotting.

Side Effects, Contraindications and Precautions

Immunologic

Immunologic side effects have included autoimmune dermatitis during the luteal phase of the menstrual cycle. Test results using interferon gamma release in vivo and vitro tests confirmed the diagnosis.

Local

Pain, irritation, and redness at the injection site.

Other

Other side effects have included viral infection (12%), hot flashes (11%), fatigue (8%), irritability (8%), worry (8%), night sweats (7%), earache, tooth disorder, anorexia, increased appetite, peripheral edema, edema, accidental injury, fever, abscess, lymphadenopathy, and herpes simplex, anaphylactoid reactions.

SPECIAL POPULATION

Renal Insufficiency

The safety and effectiveness in patients with renal insufficiency have not been established. Since Progesterone metabolites are excreted mainly by the kidneys, Progesterone should be administered with caution and careful monitoring in this patient population.

Hepatic Insufficiency

The safety and effectiveness in patients with hepatic insufficiency have not been established. Since Progesterone is metabolized by the liver, use in patients with liver dysfunction or disease is contraindicated.

Nursing Mothers

Detectable amounts of drug have been identified in the milk of mothers receiving progestational drugs.[45-47] The effects of this on the nursing infant has not been determined.

Pediatric

Not for use prior to menarche as safety and efficacy has not been established.

Surgical Patients

Whenever possible, progestins in combination with estrogens should be discontinued at least 4 to 6 weeks prior to and for 2 weeks following elective surgery associated with an increased risk of thromboembolism or during periods of prolonged immobilization.

DRUG INTERACTIONS

The metabolism of Progesterone by human liver microsomes was inhibited by ketoconazole (IC50 < 01 µM). Ketoconazole is a known inhibitor of cytochrome P450 3A4 and these data suggest that ketoconazole or other known inhibitors of this enzyme may increase the bioavailability of Progesterone. The clinical relevance of the *in vitro* findings is unknown.

CONTRAINDICATIONS[43-47]

- Undiagnosed vaginal bleeding.
- Known or suspected breast or genital cancer; history of breast cancer.
- Active deep vein thrombosis (DVT) or pulmonary embolism; history of DVT or pulmonary embolism.
- Active or recent arterial thromboembolic disease (e.g. stroke, myocardial infraction (MI), thrombophlebitis, or cerebral apoplexy.
- History of or known or suspected carcinoma of the breast or genital organs.
- Liver disease or dysfunction.
- Missed abortion or ectopic pregnancy.
- Use as a pregnancy test.
- Progesterone capsules are contraindicated in pregnant women and those with suspected pregnancy. Not effective for any purpose during pregnancy.
- Known hypersensitivity to progesterone or any ingredient in the formulation.
- *Peanut hypersensitivity*: Known hypersensitivity to peanuts because the Progesterone capsules contain peanut oil.

PRECAUTIONS

General Precautions[43-45,47]

Pretreatment General Physical Examination

A thorough physical examination prior to initiation of therapy is always advisable. The pretreatment physical examination should include special reference to breast and pelvic organs, as well as a Papanicolaou smear with relevant laboratory tests.

Fluid Retention

Because progestational drugs may cause some degree of fluid retention they should be used with caution and careful monitoring in patients with conditions (e.g. asthma, epilepsy and other seizure disorders, migraine headache, cardiac or renal impairment) that might be aggravated by fluid retention.

Bleeding Irregularities

In cases of breakthrough bleeding or irregular vaginal bleeding. Nonfunctional cases should be borne in mind and adequate diagnostic measures undertaken in such patients.

Depression

Patients who have a history of psychic depression should be carefully observed. Exercise caution in women with a history of depression; discontinue if severe depression recurs during use.

Menopause

The age of the patient constitutes no absolute limiting factor although treatment with progestin may mask the onset of the climacteric.

Diabetes Mellitus

A decrease in glucose tolerance has been observed in a small percentage of patients on estrogen-progestin combination treatment. The mechanism of this decrease is obscure. For this reason, diabetic patients should be carefully observed while receiving such therapy.

Lipid Metabolism

There are possible risks which may be associated with the use of progestin treatment, including adverse effects on lipid metabolism. The dosage used may be important in minimizing these adverse effects in patients of hyperlipidemia.

Estrogen Therapy

When used in combination with an estrogen, consider the cautions, precautions, and contraindications associated with estrogens.

CNS Effects

Patients should be warned that progesterone might cause transient dizziness or drowsiness during initial therapy. Use caution when driving or operating machinery.

Extreme dizziness and/or drowsiness, blurred vision, slurred speech, difficulty walking, loss of consciousness, vertigo, confusion, disorientation, and shortness of breath reported; women should be advised that when they experience such adverse effects, they should immediately consult their healthcare provider.

Geriatric Use

There is insufficient experience in those ≥65 years of age to determine whether they respond differently than younger adults. Select dose with caution in such patients; start at the lower end of the dosing range due to the greater frequency of decreased hepatic, renal, and/or cardiac function and of concomitant disease and drug therapy.

Hepatic Impairment

If indicated in patients with mild to moderate hepatic dysfunction, monitor patient closely.

Renal Impairment

Progesterone injection and capsules: Use with caution; carefully monitor patients with renal impairment.

Lactation

Progestins are distributed into milk. Hence caution is advised in lactating women. Detectable amounts of drug have been identified in the milk of nursing mothers receiving progestational drugs. The effect of this on the nursing infant has not been determined.

Pathological Examination

The pathologist should be advised of progestin therapy when relevant specimens are submitted.

Prolonged Therapy

Effect of long-term therapy on pituitary, ovarian, adrenal, hepatic, or uterine function has not been determined.

Ocular Effects

If unexplained, sudden or gradual, partial or complete loss of vision; proptosis or diplopia; papilledema; migraine; or retinal vascular lesions occur, discontinue and initiate appropriate diagnostic and therapeutic measures. Discontinue permanently and do *not* reinitiate therapy if ocular examination reveals evidence of papilledema or retinal vascular thrombosis or lesions.

Carcinogenesis and Mutagenesis—Carcinoma of Breast

Mitotic activity in the breast is maximum in the luteal phase of the menstrual cycle, so it was postulated that progesterone is responsible for breast tumors. Though progesterone is the primary stimulus for mammary growth and differentiation, with increasing exposure it limits breast epithelial growth as it does in endometrial epithelium. It inhibits estrogen-induced proliferation. Most studies show that high levels of estrogen and progesterone do not adversely impact the course of breast carcinoma.

A small increase in the risk of breast cancer is associated with exogenous estrogen-progestin HRT in postmenopausal women. It is not clear whether it initiates new cancers or impacts undiagnosed pre-existing tumors. Women with a greater breast density have been shown to have a higher risk of cancer with the risk increasing 4–5 times if more than 75% of the breast is dense as the density reflects epithelial and stromal cell proliferation. Breast density has a strong genetic component as shown by studies in families and twins. This genetic influence is shared with other genetic factors which increase the risk of breast cancer. Breast density decreases with increasing age, number of pregnancies and body weight.

Based on data from the Women's Health Initiative (WHI) studies, an increased risk of invasive breast cancer was observed in postmenopausal women using conjugated estrogens (CE) in combination with medroxyprogesterone acetate (MPA). This risk may be associated with duration of use and declines once combined therapy is discontinued. The risk of invasive breast cancer was decreased in postmenopausal women with a hysterectomy using CE only, regardless of weight. However, the risk was not significantly decreased in women at high risk for breast cancer (family history of breast cancer, personal history of benign breast disease). An increase in abnormal mammogram findings has also been reported with estrogen alone or in combination with progestin therapy. Use is contraindicated in patients with known or suspected breast cancer. The breast cancer produced by EPT is well differentiated, node negative and is diagnosed at an early stage.

Long-term intramuscular administration of MPA has been shown to produce mammary tumors in beagle dogs. There is no evidence of a carcinogenic effect associated with the oral administration of MPA to rats and mice. It is not associated with an increased risk of breast cancer when used for contraception over a long duration. Medroxyprogesterone acetate was not mutagenic in a battery of *in vitro* or *in vivo* genetic toxicity assays.

Impairment of Fertility

Progesterone at high doses is an antifertility drug and high doses would be expected to impair fertility until the cessation of treatment.

Cardiovascular or Cerebrovascular Disorders[43-47]

Possible cardiac disorders (MI) or thromboembolic and thrombotic disorders (e.g. thrombophlebitis, pulmonary embolism, cerebrovascular disorders, retinal thrombosis). Observe patients for these effects; discontinue immediately and do not re-administer if these disorders occur or are suspected.

Estrogens with or without progestin should not be used to prevent cardiovascular disease. Using data from the WHI studies, an increased risk of DVT and stroke has been reported with CE and an increased risk of DVT, stroke, pulmonary emboli (PE) and MI has been reported with CE with MPA in postmenopausal women (50-79 years) during 5.6 years of study. Additional risk factors include diabetes mellitus, hypercholesterolemia, hypertension, SLE, obesity, tobacco use, and/or history of venous thromboembolism (VTE). Risk factors should be managed appropriately; discontinue use if adverse cardiovascular events occur or are suspected.

Dementia

Estrogens with or without progestin should not be used to prevent dementia. In the Women's Health Initiative Memory Study (WHIMS), an increased incidence of dementia was observed in women ≥ 65 years of age taking CE alone or in combination with MPA. It is unknown whether the same effects are seen in younger postmenopausal women.

Endometrial Cancer

Progestins are used to reduce the risk of endometrial hyperplasia in nonhysterectomized postmenopausal women receiving conjugated estrogens. The use of unopposed estrogen in women with an intact uterus is associated with an increased risk of endometrial cancer. The addition of a progestin to estrogen therapy may decrease the risk of endometrial hyperplasia, a precursor to endometrial cancer. Adequate diagnostic measures, including endometrial sampling if indicated, should be performed to rule out malignancy in postmenopausal women with undiagnosed abnormal vaginal bleeding. Estrogens may exacerbate endometriosis. Malignant transformation of residual endometrial implants has been reported posthysterectomy with unopposed estrogen therapy. Consider adding a progestin in women with residual endometriosis posthysterectomy.

Chapter 13

Patient Advice

- Importance of advising patients of anticipated menstrual effects.
- *Progesterone capsules*: Importance of using caution when driving or operating machinery, especially during the initiation of therapy.
- *Progesterone vaginal gel*: Importance of following special instructions if the gel is used at altitudes >2500 feet.
- Importance of discontinuing therapy and contacting clinician if sudden partial or complete vision loss, blurred vision, bulging of the eyes, double vision, or migraine occur.
- Importance of discontinuing therapy and contacting clinician if any symptoms of thromboembolic and thrombotic disorders occurs.
- Importance of women informing clinicians if they are or plan to become pregnant or plan to breastfeed.
- Importance of informing clinicians of existing or contemplated concomitant therapy, including prescription and over-the-counter (OTC) drugs, as well as concomitant illnesses.
- Advise women using the vaginal gel or insert not to use other vaginal preparations. If therapy with another agent administered intravaginally is needed in a woman using the gel, administer the other agent 6 hours before or after progesterone vaginal gel.
- Importance of informing patients of other important precautionary information (*see* Chapter 12 for Cautions).

Chapter 14

Drug and Test Interactions

There are several notable drug interactions with progesterone.
- Certain selective serotonin reuptake inhibitors (SSRIs) may increase the GABA$_A$ receptor-related central depressant effects of progesterone by enhancing its conversion into 5α-dihydroprogesterone and allopregnanolone via activation of 3α-hydroxysteroid dehydrogenase (3α-HSD).[62]
- Progesterone potentiates the sedative effects of benzodiazepines and ethanol. Notably, there is a case report of progesterone abuse alone with very high doses.[63,64]
- 5α-reductase inhibitors, such as finasteride and dutasteride, as well as inhibitors of 3α-HSD, such as medroxyprogesterone acetate, inhibit the conversion of progesterone into its inhibitory neurosteroid metabolites, and for this reason, may have the potential to block or reduce its sedative effects.[43,44,65]
- Progesterone is a weak but significant agonist of the "pregnane X receptor" (PXR), and has been found to induce several hepatic cytochrome P450 enzymes, such as CYP3A4, especially when concentrations are high, such as with pregnancy range levels.[67-70] As such, progesterone may have the potential to accelerate the clearance of various drugs, especially with oral administration (which results in supraphysiological levels of progesterone in the liver), as well as with the high concentrations achieved with sufficient injection dosages.
- *Anticoagulants*: Progestins may diminish the therapeutic effect of anticoagulants. More specifically, the potential prothrombotic effects of some progestins and progestin-estrogen combinations may counteract anticoagulant effects. *Management*: Carefully weigh the prospective benefits of progestins against the potential increased risk of procoagulant effects and thromboembolism. Use is considered contraindicated under some circumstances. Refer to related guidelines for specific recommendations. *Consider therapy modification.*

- *Antifungal agents (vaginal)*: May diminish the therapeutic effect of Progesterone. *Avoid combination.*
- *C1 inhibitors*: Progestins may enhance the thrombogenic effect of C1 inhibitors. *Monitor therapy.*
- *Herbs (Progestogenic properties, e.g. Bloodroot, Yucca)*: May enhance the adverse/toxic effect of Progestins. *Monitor therapy.*
- *Pomalidomide*: Progestins may enhance the thrombogenic effect of pomalidomide. *Management*: Canadian pomalidomide labeling recommends caution with use of hormone replacement therapy and states that hormonal contraceptives are not recommended. US pomalidomide labeling does not contain these specific recommendations. *Consider therapy modification.*
- *Ulipristal*: May diminish the therapeutic effect of Progestins. Progestins may diminish the therapeutic effect of ulipristal. *Management*: Ulipristal for uterine fibroids (Canadian indication): avoid progestins within 12 days of stopping ulipristal; as emergency contraceptive (US indication): avoid progestins within 5 days of stopping ulipristal. *Avoid combination.*

TEST INTERACTIONS

Thyroid function, metyrapone, liver function, coagulation tests, endocrine function tests are altered with concomitant progestin therapy.

The following laboratory results may be altered by the use of estrogen-progestin combination drugs:
- Increased sulfobromophthalein retention and other hepatic function tests
- *Coagulation tests*: Increase in prothrombin factors VII, VIII, IX, and X
- Metyrapone test
- Pregnanediol determination
- *Thyroid function*: Increase in protein-bound iodine (PBI) and butanol extractable protein bound iodine, and decrease in T3 uptake values.

Chapter 15

Prescription Writing

PROGESTIN CHALLENGE TEST

First Clinical Step used in patient with amenorrhea (primary/secondary/lactational or perimenopausal). If positive, confirms intact HPO axis, endogenous estrogen, receptive endometrium and patent outflow.

Regimes
1. Oral: Micronized progesterone: 300 mg OD for 5 days.
2. Tab. Medroxyprogesterone acetate (MPA) 10 mg OD for 7-10 days.
3. Dydrogesterone-10 mg once a day for 10 days
4. Progesterone 4% vaginal gel: One prefilled applicator every other day for a total of 6 doses.
5. Parenteral-Progesterone in oil: 100-200 mg IM single dose.

Postponement of Menstruation

Norethisterone 5 mg TDS/Tab MPA 10 mg BD at least 3 days prior to the expected period date and continued till desire.

Minipill (Progestin-only Pill)

It contains small doses of progestin, e.g.
1. Cerazette: 0.075 mg Desogestrel
2. Exluton: 0.5 mg Lynestrenol
3. Ovrette: 0.075 mg Norgestrel
4. Femulen: 0.5 mg Ethynodiol diacetate
5. Microval: 0.03 mg Levonorgestrel.

Dose: It is started on first day of the menses taken daily in continuous fashion at the same time of the day, preferably not more than 3 hours gap.

If one pill is missed, back-up contraception should be taken for 24 hours.

EMERGENCY CONTRACEPTION

a. Tab. Levonorgestrel: 1.5 mg single dose orally within 72 hours
b. Ulipristal acetate 30 mg Orally single dose within 120 hours of intercourse
c. Tab. Ethinylestradiol 0.2 mg + Tab. Levonorgestrel 2 mg orally 2 tab and repeated after 12 hours
d. Tab. Ethinylestradiol (50 microgram) + Tab. Norgestrel (0.25 mg) orally 2 tab and repeated after 12 hours
e. Cu-T insertion, within 5 days—most effective method
f. Tab. Mifepristone 25–50 mg single dose within 120 hours of exposure.

ABNORMAL UTERINE BLEEDING (AUB)

1. Tab. Tranexamic acid 500–1000 mg 3–4 times a day × 5 days
2. Tab. Mefenamic acid 500 mg 3 times a day × 5 days
3. Tab. Norethisterone 10 mg 3 times a day till bleeding stop
   - Followed by 10 mg 2 times a day × 7 days
   - Followed by 10 mg once daily × 14 days for first cycle
   - Followed by Tab. Norethisterone 10 mg once daily D5-D26 for 3–6 cycles.
   Or
   - Tab. Medroxyprogesterone 10 mg BD from D5-D25 (3–6 months)
4. Oral contraceptive pills (containing ethinylestradiol 35 µg) BD to QID × 5 days
   tapering to 1 BD × 28 days for 3–6 cycles
5. Intrauterine progesterone (LNG-IUS) over 5 years
6. Depot medroxyprogesterone acetate (DMPA) intramuscular
7. Danazol 200–800 mg daily into two or four divided doses
8. Surgical management (non-responder to medical management or family completed).

Endometriosis

1. Nonsteroidal anti-inflammatory drugs (NSAIDs)
2. Progestogens
   - Tab. Medroxyprogesterone acetate 30 mg OD × 3–6 months
   - Tab. Megestrol acetate 40 mg OD
   - Tab. Lynestrenol 10 mg OD
   - Tab. Dydrogesterone 20–30 mg OD
   - Tab. Dienogest 2 mg orally once daily
   - Mirena as intrauterine contraceptive device (IUCD)

3. Antiprogestins
   - Tab. Gestrinone: 1.25 mg-2.5 mg twice weekly × 6-9 months
   - Tab. Danazol 400-800 mg orally in four divided doses × 6-9 months
4. GnRH analogs (Medical Oophorectomy)
   - Inj. Leuprolide: 3.75 mg IM monthly or 500 μg S/C once daily
   - Inj. Goserelin 3.6 mg S/C monthly
   - Inj. Nafarelin 200 mg twice daily as intranasal spray.

FIBROID UTERUS

1. Tab. Tranexamic acid 500-1000 mg 3-4 times a day × 5 days (during menstruation)
2. Leuprolide acetate depot (3.75 mg/month or 11.25 mg/3 months intramuscularly)
   - Inj. Goserelin (3.6 mg every 28 days for 3-6 months subcutaneously)
   - Nafarelin 800 μg intranasal in 2-3 divided doses
3. Tab. Ulipristal acetate 5 or 10 mg/day orally × 3-6 months
4. Tab. Mifepristone 5-50 mg/day orally 3-6 months
5. Intrauterine progestestin (LNG-IUS) over 5 years
6. Tab. Danazol 200-800 mg daily into two or four divided doses for 6-12 months
7. Focused ultrasound treatment/Surgical management (Non-responder to medical management).

HORMONE THERAPY FOR MENOPAUSE

Regimens	Estrogens	Progestins
Cyclic sequential	D 1-25	D 13-25
Continuous sequential	Daily	D 1-14
Continuous combined	Daily	Daily
Cyclic combined	D 1-25	D 1-25

Estrogens : 0.625 mg conjugated estrogen
2 mg Estradiol valerate
1 mg Micronized Estradiol
50 μg transdermal Estradiol

Progestins : 200 mg micronized progesterone
(Two weeks   10 mg MPA
every month) 10 mg dydrogesterone

(Daily) : 0.35 to 1 mg norethindrone
100 mg Micronized progesterone
2.5 mg MPA
1. In hysterectomized patients
   - Tab. conjugated equine estrogen (CEE) 0.625 mg OD or estradiol valerate 1-2 mg OD
   - Along with calcium carbonate 1000 mg and bisphosphonate 35 mg weekly empty stomach
2. In natural menopause
   - CEE 0.625 mg and medroxyprogesterone acetate 2.5 mg daily
   - 17β estradiol 1/2 mg and dydrogesterone 5/10 mg in continuous combined or continuous sequential manner
3. If HRT is contraindicated then
   - SERM 60 mg (Raloxifene)
   Or
   - Tibolone 2.5 mg 1 OD can be given.

ENDOMETRIAL CANCER

Progestins are the most commonly used adjuvant or palliative therapy in early/advanced or recurrent endometrial cancer.

Medroxyprogesterone acetate (MPA) and Megestrol acetate are the most commonly used agents. These drugs are continued for at least 3 months and remain an important alternative for women who are unable to tolerate chemotherapy.

Regimes

- Medroxyprogesterone acetate 1 gm week or 150 mg/day orally
- Megestrol acetate 160 mg/day orally
- 17 Hydroxyprogesterone caproate 1 gm/week IM.

LUTEAL PHASE SUPPORT IN ASSISTED REPRODUCTIVE TECHNOLOGY

Used to support luteal phase, embryo implantation and early pregnancy as part of assisted reproductive technology (ART) treatment of infertile women. It can be given as:
- Progesterone in oil 50 mg im daily
- Vaginal suppository 200 mg twice daily
- Micronized progesterone orally 200 mg twice daily usually continued for 14 days

- *Progesterone 8% vaginal gel*: For women, who require progesterone supplementation, administer contents of one prefilled applicator once daily. If pregnancy occurs, continue until placenta autonomy is achieved, up to 10–12 weeks.
- *Progesterone vaginal insert*: One insert (100 mg) 2–3 times daily. Start at oocyte retrieval; continue for up to 10 weeks.

Preterm Labor

For prevention of preterm labor (PTL) in women with history of previous PTL:
- 17α-hydroxyprogesterone 250 mg IM weekly
- Micronized progesterone 100 mg daily vaginally.

For prevention of PTL in women with short cervix of <15 mm detected on transvaginal ultrasound at 22–26 weeks:
- Micronized progesterone 200 mg daily vaginally
- Progesterone vaginal insert 100 mg daily.

To be started from 16 to 24 weeks and continued until 36–37 weeks is recommended.

Threatened and Recurrent Abortions

Threatened Miscarriage

Oral route: Dydrogesterone 40 mg loading dose followed by 20–30 mg daily till 7 days after bleeding (FOGSI).

Vaginal route: Micronized progesterone 400 mg/day vaginally till bleeding stops (FOGSI).

Recurrent Miscarriage

Oral route: Dydrogesterone 10 mg BD till 20 weeks of pregnancy

Vaginal route: Micronized progesterone: 400 mg/day vaginally till 20 weeks of pregnancy.

Note: Oral MCP is not indicated for threatened and/or recurrent miscarriage.

Chapter 16

References

1. Marshall WJ, Bangert SK. Clinical Chemistry. Elsevier Health Sciences; 2008.
2. Yamazaki H, Shimada T. Progesterone and testosterone hydroxylation by cytochromes P450 2C19, 2C9, and 3A4 in human liver microsomes. Arch Biochem Biophys. 1997;346(1):161-9.
3. Stanczyk FZ. Pharmacokinetics and potency of progestins used for hormone replacement therapy and contraception. Rev Endocr Metab Disord. 2002;3(3):211-24.
4. Simon JA, Robinson DE, Andrews MC, et al. The absorption of oral micronized progesterone: the effect of food, dose proportionality, and comparison with intramuscular progesterone. Fertil Steril. 1993;60(1): 26-33.
5. Speroff L, Fritz MA. Clinical Gynecologic Endocrinology and Infertility. Philadelphia: Lippincott Williams & Wilkins; 2012.
6. McKay GA, Walters MR. Lecture Notes: Clinical Pharmacology and Therapeutics. John Wiley & Sons; 2013, p. 33.
7. Zutshi. Hormones in Obstetrics and Gynaecology. Jaypee Brothers Publishers; 2005, p. 74.
8. Cometti B. Pharmaceutical and clinical development of a novel progesterone formulation. Acta Obstetricia Et Gynecologica Scandinavica. 2015;94(Suppl 161): 28-37.
9. Adler N, Pfaff D, Goy RW. Handbook of Behavioral Neurobiology, Volume 7 Reproduction (1st ed.). New York: Plenum Press; 2012, p. 189.
10. Progesterone (CHEBI:17026). ChEBI. European Molecular Biology Laboratory-EBI. Retrieved July 4, 2015.
11. King TL, Brucker MC. Pharmacology for Women's Health. Jones & Bartlett Publishers; 2010, pp. 372-3.
12. Baulieu E, Schumacher M. Progesterone as a neuroactive neurosteroid, with special reference to the effect of progesterone on myelination. Steroids. 2000;65(10-11):605-12.
13. 19th WHO Model List of Essential Medicines (April 2015) (PDF). WHO. April 2015. Retrieved May 10, 2015.
14. Josimovich J. Gynecologic Endocrinology. Springer Science & Business Media; 2013, pp. 9, 25-9.
15. Ruan X, Mueck AO. Systemic progesterone therapy-oral, vaginal, injections and even transdermal? Maturitas. 2014;79(3):248-55.

References

16. Gerald, M. The Drug Book. New York, NY: Sterling Publishing; 2013, p. 186.
17. Sauer MV. Principles of Oocyte and Embryo Donation. Springer Science & Business Media; 2013, pp. 7, 118.
18. Minkin MJ, Wright CV. A Woman's Guide to Menopause & Perimenopause. Yale University Press; 2005.
19. Racowsky C, Schlegel PN, Fauser BC, et al. Biennial Review of Infertility. Springer Science & Business Media; 2011, pp. 84-5.
20. Index Nominum 2000: International Drug Directory. Taylor & Francis. January 2000. p. 880.
21. Orange Book: Approved Drug Products with Therapeutic Equivalence Evaluations: 020701. Food and Drug Administration. 2010-07-02. Retrieved 2010-07-07.
22. Lark S. Making the Estrogen Decision. McGraw-Hill Professional; 1999, p. 22.
23. Zava DT, Dollbaum CM, Blen M (March 1998). Estrogen and progestin bioactivity of foods, herbs, and spices. Proceedings of the Society for Experimental Biology and Medicine. Proc Soc Exp Biol Med. 1998;217(3): 369-78.
24. Orange Book: Approved Drug Products with Therapeutic Equivalence Evaluations: 022057. Food and Drug Administration. 2010-07-02. Retrieved 2010-07-07.
25. Komesaroff PA, Black CV, Cable V, et al. Effects of wild yam extract on menopausal symptoms, lipids and sex hormones in healthy menopausal women. Climacteric. 2001;4(2):144-50.
26. Orange Book: Approved Drug Products with Therapeutic Equivalence Evaluations: 075906. Food and Drug Administration. 2010-07-02. Retrieved 2010-07-07.
27. da Fonseca EB, Bittar RE, Carvalho MH, et al. Prophylactic administration of progesterone by vaginal suppository to reduce the incidence of spontaneous preterm birth in women at increased risk: a randomized placebo-controlled double-blind study. Am J Obstet Gynecol. 2003; 188(2):419-24.
28. Harris, Gardiner. Hormone Is Said to Cut Risk of Premature Birth. New York Times. May 2, 2011. Retrieved May 5, 2011.
29. O'Brien JM, Adair CD, Lewis DF, et al. Progesterone vaginal gel for the reduction of recurrent preterm birth: primary results from a randomized, double-blind, placebo-controlled trial. Ultrasound Obstet Gynecol. 2007;30(5):687-96.
30. DeFranco EA, O'Brien JM, Adair CD, et al. Vaginal progesterone is associated with a decrease in risk for early preterm birth and improved neonatal outcome in women with a short cervix: a secondary analysis from a randomized, double-blind, placebo-controlled trial. Ultrasound Obstet Gynecol. 2007;30(5):697-705.
31. Fonseca EB, Celik E, Parra M, et al. Progesterone and the risk of preterm birth among women with a short cervix. N Engl J Med. 2007;357(5):462-9.

32. Romero R. Prevention of spontaneous preterm birth: the role of sonographic cervical length in identifying patients who may benefit from progesterone treatment. Ultrasound Obstet Gynecol. 2007;30(5):675-86.
33. Hassan SS, Romero R, Vidyadhari D, et al. Vaginal progesterone reduces the rate of preterm birth in women with a sonographic short cervix: a multicenter, randomized, double-blind, placebo-controlled trial. Ultrasound in Obstetrics & Gynecology. 2011;38(1):18-31.
34. Progesterone-Drugs.com, retrieved 2015-08-23.
35. Progesterone helps cut risk of pre-term birth. Women's health. msnbc.com. 2011-12-14. Retrieved 2011-12-14.
36. Devinsky O, Schachter S, Pacia S. Complementary and Alternative Therapies for Epilepsy. Demos Medical Publishing; 2005.
37. Sriram D. Medicinal Chemistry. New Delhi: Dorling Kindersley India Pvt. Ltd.; 2007, p. 432.
38. Heyl FW. Progesterone from 3-Acetoxybisnor-5-cholenaldehyde and 3-Ketobisnor-4-cholenaldehyde. J Am Chem Soc. 1950;72(6):2617-9.
39. Johnson WS, Gravestock MB, McCarry BE. Acetylenic bond participation in biogenetic-like olefinic cyclizations. II. Synthesis of dl-progesterone. J Am Chem Soc. 1971;93(17):4332-4.
40. Coutinho EM, Segal SJ. Is Menstruation Obsolete? New York: Oxford University Press; 1999.
41. Niño J, Jiménez DA, Mosquera OM, et al. Diosgenin quantification by HPLC in a Dioscorea polygonoides tuber collection from colombian flora. J Braz Chem Soc. 2007;18(5):1073-6.
42. Wright DW, Kellermann AL, Hertzberg VS, et al. ProTECT: a randomized clinical trial of progesterone for acute traumatic brain injury. Ann Emerg Med. 2007;49(4):391-402, 02.e1-2.
43. Meyer L, Venard C, Schaeffer V, et al. The biological activity of 3alpha-hydroxysteroid oxido-reductase in the spinal cord regulates thermal and mechanical pain thresholds after sciatic nerve injury. Neurobiol Dis. 2008;30(1):30-41.
44. Pazol K, Wilson ME, Wallen K. Medroxyprogesterone acetate antagonizes the effects of estrogen treatment on social and sexual behavior in female macaques. J Clin Endocrinol Metab. 2004;89(6):2998-3006.
45. Schindler AE, Campagnoli C, Druckmann R, et al. Classification and pharmacology of progestins. Maturitas. 2003;46(Suppl 1): S7-S16.
46. Simon JA, Robinson DE, Andrews MC, et al. The absorption of oral micronized progesterone: the effect of food, dose proportionality, and comparison with intramuscular progesterone. Fertil Steril. 1993;60(1):26-33.
47. Rakel D. Integrative Medicine. Philadelphia: Elsevier Health Sciences; 2012, p. 343.
48. Noguchi E, Fujiwara Y, Matsushita S, et al. Metabolism of tomato steroidal glycosides in humans. Chem Pharm Bull. 2006;54(9):1312-4.

49. Yang DJ, Lu TJ, Hwang LS. Isolation and identification of steroidal saponins in Taiwanese yam cultivar (Dioscorea pseudojaponica Yamamoto). J Agric Food Chem. 2003;51(22):6438-44.
50. Hooker E. Final report of the amended safety assessment of Dioscorea Villosa (Wild Yam) root extract. Int J Toxicol. 2004;23(Suppl 2):49-54.
51. Myoda T, Nagai T, Nagashima T. Properties of starches in yam (Dioscorea spp.) tuber. Current Topics in Food Science and Technology; 2005, pp. 105-14.
52. Bewick PM. Medicinal natural products: a biosynthetic approach. New York: Wiley; 2002, p. 244.
53. Reddy DS. Neurosteroids: endogenous role in the human brain and therapeutic potentials. Progress in Brain Research. 2010;186:113-7.
54. Wang-Cheng R, Neuner JM, Barnabei VM. Menopause. USA: ACP Press; 2007, p. 97.
55. de Wit H, Schmitt L, Purdy R, et al. Effects of acute progesterone administration in healthy postmenopausal women and normally-cycling women. Psychoneuroendocrinology. 2001;26(7):697-710.
56. van Broekhoven F, Bäckström T, Verkes RJ. Oral progesterone decreases saccadic eye velocity and increases sedation in women. Psychoneuroendocrinology. 2006;31(10):1190-9.
57. Arun N, Narendra M, Shikha S. Progress in Obstetrics and Gynecology—3. New Delhi: Jaypee Brothers Medical Publishers Pvt. Ltd; 2012.
58. Unfer V, Casini ML, Marelli G, et al. Different routes of progesterone administration and polycystic ovary syndrome: a review of the literature. Gynecol Endocrinol. 2005;21(2):119-27.
59. Elshafie MA, Ewies AA (2007). Transdermal natural progesterone cream for postmenopausal women: inconsistent data and complex pharmacokinetics. J Obstet Gynaecol. 2007;27(7):655-9.
60. Du JY, Sanchez P, Kim L, et al. Percutaneous progesterone delivery via cream or gel application in postmenopausal women: a randomized cross-over study of progesterone levels in serum, whole blood, saliva, and capillary blood. Menopause (New York, NY). 2013;20(11):1169-75.
61. World Professional Association for Transgender Health (September 2011), Standards of Care for the Health of Transsexual, Transgender, and Gender Nonconforming People, Seventh Version (PDF).
62. Pinna G, Agis-Balboa RC, Pibiri F, et al. Neurosteroid biosynthesis regulates sexually dimorphic fear and aggressive behavior in mice. Neurochem Res. 2008;33(10):1990-2007.
63. Babalonis S, Lile JA, Martin CA, et al. Physiological doses of progesterone potentiate the effects of triazolam in healthy, premenopausal women. J. Psychopharmacol. 2011;215(3):429-39.
64. "Progesterone abuse". Reactions Weekly. Springer International Publishing. 1996;599(1):9.
65. Traish AM, Mulgaonkar A, Giordano N. The dark side of 5α-reductase inhibitors' therapy: sexual dysfunction, high Gleason grade prostate cancer and depression. Korean J Urol. 2014;55(6):367-79.

66. Deshmukh S. Infertility Management Made Easy. New Delhi: Jaypee Brothers Medical Publishers Pvt Ltd; 2013, p. 273.
67. Meanwell NA. Tactics in Contemporary Drug Design. Berlin: Springer; 2014.
68. Legato MJ, Bilezikian JP. Principles of Gender-specific Medicine. Gulf Professional Publishing; 2004.
69. Lemke TL, Williams DA. Foye's Principles of Medicinal Chemistry. Lippincott Williams & Wilkins; 2012.
70. Estrogens—Advances in Research and Application: 2013 Edition: Scholarly Brief. Scholarly Editions. June 21, 2013. ISBN: 978-1-4816-7550-5.

Index

Note: Page numbers followed by *t* refer to table

A

Ablation, endometrial 51
Abortion
    missed 70
    recurrent 60, 83
    threatened 60, 83
Abscess 69
Absorption 17
Accidental injury 69
Acetate 41
Acne 68
Adenomyosis 49
Adrenal hyperplasia 19
Allylestrenol 10, 58
Alopecia 68
Alzheimer's disease 26
Amenorrhea 35, 49, 68
    lactational 35
    primary 35
    secondary 35
Angina 48
    pectoris 67
Anorexia 69
Anti-androgenic action 34
Antifungal agents 78
Antiprogestins 61, 81
Arthritis 68
Artificial heart valves 48
Assisted reproductive technology 35, 55, 61, 82
Asthenia 67
Asthma 71

B

Benzyl alcohol 6
Biopsy, endometrial 54
Biphasic pill 63
Bleeding, abdominal 54
Bloating 67
Blood
    test 11
    vessels, coiling of 24
Breast 36
    cancer 23, 48
        risk, reduction of 51
    carcinoma 50, 67, 73
    diseases, benign 65
    enlargement 67
    pain 67
    tenderness 65, 67
Breastfeeding 36
Bronchitis 68

C

Cancer, endometrial 23, 35, 54, 75, 82
Carbohydrate metabolism 25, 26
Carcinogenesis 73
Carcinoma 36
Cardiac disorders 74
Cardiovascular disease 48
Catamenial epilepsy 61
Cells, endometrial 24
Cerebral
    apoplexy 70
    edema 67
    thrombosis 67
Cerebrovascular disorders 74
Cholecystectomy 68
Cholesterol 6
Colorectal cancer, reduction of 65
Combined contraceptive vaginal rings 66
Combined hormonal
    contraception 62

non-oral method 65
oral methods 64
Combined injectable contraceptives 66
Combined oral contraceptive pills 33, 36, 37, 63
Confusion 67
Congenital adrenal hyperplasia 11
Constipation 67, 68
Contraception, long-acting 43
Contraceptive 36
　implants 43, 44t, 47
　methods 41
　subdermal implants 44
Coronary artery disease 48
Corpus luteum 11
Cough 68
Cyclic estrogen-progestin contraceptive 53
Cyproterone acetate 9, 34
Cystitis 67

D

Deep vein thrombosis 70
Dementia 75
Depot medroxyprogesterone acetate 28, 53
Depression 67, 71
Desogestrel 10, 12, 20, 25, 30, 34, 37, 41, 42, 63
Diabetes mellitus 48, 71, 75
Diarrhea 67, 68
Dienogest 33, 34
Dihydroxyprogesterone 22
Dizziness 67
Drospirenone 9, 33, 34, 63
Dry mouth 67
Dydrogesterone 9, 12, 27, 35, 51, 79
Dysmenorrhea 49
Dyspareunia 49
Dyspepsia 67
Dysuria 67

E

Earache 69
Edema 69
Emergency contraceptive pill 42

Endocrine function tests 78
Endometrial
　cancer
　　prevention of 50
　　reduction of 65
　carcinoma
　　primary 28
　　secondary 28
　hyperplasia, treatment of 50
Endometriosis 35, 49, 51, 57, 65, 80
Endometrium 24
　transformation of 25
Epilepsy 71
Erythema
　multiforme 68
　nodosum 68
Estradiol valerate 51
Estranes 10, 30
Estrogen 50, 63, 72, 75
　contraceptives 48
　therapy 72
Ethinyl estradiol 33, 64
Ethynodiol diacetate 41

F

Family planning 29, 35
Fatigue 67, 69
Fertility, impairment of 74
Fever 69
Fibrinolysis 52
Fibroid 49
　uterus 81
Fluid retention 71
Follicular stimulating hormone 62
Fungal vaginitis 67

G

Galactorrhea 68
Gallbladder disease 48
Gastroenteritis 67
Gastrointestinal metabolism 18
Genital bleeding 48
Gestodene 10, 12, 20, 25, 31, 34
Glucose intolerance 68
Gonadotropin-releasing hormone 55
Gonanes 10, 30

H

Headache 67, 71
Hemorrhagic eruption 68
Hepatic
 impairment 72
 insufficiency 69
 metabolism 18
Hepatitis 68
Hiatus hernia 68
Hidradenitis suppurativa 61
Hirsutism 68
Hormonal contraceptives 62
Hormone
 luteinizing 21, 42, 62
 replacement therapy 15, 16, 28, 35, 49, 50
Hot flashes 69
Hypercholesterolemia 75
Hyperlipidemia 72
Hyperplasia, endometrial 35, 50, 75
Hypertension 48, 67, 75
Hypertonia 68
Hypothalamic-pituitary axis 53
 suppression of 24
Hysterectomy, supracervical 51
Hysteroscopy 54

I

Infertility 35
Insomnia 67
Intrauterine device 41
Itching 68

J

Jaundice, cholestatic 68

K

Kypothalamus, level of 24

L

Lactation 42, 72
Leg cramps 68

Leucorrhea 67
Levonorgestrel 10, 12, 20, 31, 34, 37, 41, 42
Lipid metabolism 24
Lipoprotein, high-density 25
Liver
 disease 48, 70
 function 78
 tumors, benign 48
Low density lipoprotein 22
Luteal phase defect 16, 35, 55
Lymphadenopathy 69
Lynestrenol 10, 41

M

Malignant liver tumors 48
Medroxyprogesterone acetate 9, 12, 20, 25, 28, 34, 36, 55, 66, 74, 79
Megestrol acetate 9, 55
Menopause 71, 81
Menorrhagia 49
Menstrual cycle, luteal phase of 24
Menstruation, postponement of 57, 79
Metyrapone 78
Micronization 15
Migraine 67, 68, 71
Minipill 79
Mirena 41, 48
Miscarriage
 recurrent 83
 threatened 83
Monophasic pill 63
Multiple sclerosis 61
Muscle disorder 68
Mutagenesis 73
Myalgia 68
Myocardial infarction 48
Myomas 35
Myometrium 24, 58

N

Nasal congestion 68
Nausea 65, 67, 68
Nervous system 26, 67
Nestorone 9, 33, 34, 44
Neuritis, ocular 68
Nexplanon 44

Night sweats 69
Ninety-one-day oral pill 63
Nomegestrol 9, 32, 34
Noninvasive method 65
Nonsteroidal anti-inflammatory drugs 80
Norethindrone 10, 12, 20, 25, 30, 51, 64
Norethisterone 34, 37, 41
    acetate 12
Norethynodrel 10
Norgestimate 10, 12, 20, 25, 32, 34, 64
Norgestrel 25, 41
Norplant 41, 44, 45
Nuvaring 38

O

Obesity 75
One yearly pill 63
Oral contraceptive pill 31, 35, 62
    action of 62
Osteopenia 23
Osteoporosis 50, 51
Ovarian cancer, reduction of 65
Ovary
    endometrioid tumors of 51
    functional cysts of 65
Ovulation, inhibition of 25

P

Pain 6
    abdominal 67
    back 68
    chest 67
    musculoskeletal 68
Papilledema 68
Parenteral injection 14
Peak plasma concentrations 17
Peanut hypersensitivity 70
Pelvic inflammatory disease 65
Peri-implantation period 24
Perimenopausal period 35
Peripheral edema 69
Pharyngitis 68
Pneumonitis 68
Polymethylsiloxane 48
Pomalidomide 78

Pregnancy 22
    ectopic 48, 70
    test 70
Pregnane X receptor 77
Pregnanediol 18
Pregnanes 9
Pregnanetriol 18, 19
Pregnanolone 6, 18
Premenstrual syndrome 16, 34, 35, 56, 65
    correction of 65
    treatment of 49
Preterm labor 16, 58
Progesterone 1, 5, 6, 8, 11, 12, 18, 22, 23, 25, 34, 36, 58, 61, 78, 82
    capsules 70, 76
    challenge test 16, 35
    esters of 28
    gel 68
    low levels of 21
    micronized 15, 35, 51, 61, 79, 82
    natural 4, 58
    oral micronized 4, 17
    physiological actions of 21
    receptor 1
    side effects of 67
    suppositories 58
    uses of 61
    vaginal gel 13, 36, 76, 79
Progestin 17, 25t, 41, 50, 63, 78
    challenge test 79
    containing intrauterine devices 36
    only pill 30, 79
    role of 62
    therapeutic applications of 35t
    vaginal ring 66
Progestogen 1, 8, 60, 80
    classification of 9
    only contraceptives 40, 41t
    only pill 35, 36, 41, 35
    types of 64
Promegestone 9
Proptosis 68
Pruritus 68
Pyrexia 67

Q

Quadriphasic pill 63

Index

R
Rash 68
Renal impairment 72
Retinal thrombosis 68, 74

S
Scalp hair, loss of 68
Seizure disorders 71
Selective progesterone receptor
    modulators 61
Selective serotonin reuptake inhibitors 77
Sex hormone binding globulin 25, 31, 34, 37
Sexually transmitted diseases 65
Sinusitis 68
Skin discoloration 68
Speech disorder 67
Steroidogenesis 7
Stimulate prolactin secretion 24
Stratified squamous vaginal epithelium 38
Stroke 70
Stroma, decidualization of 24
Stromal cells
    endometrial 21
    predecidualization of 24
Syncope 67
Systemic disease, prevention of 23

T
Testosterone 8
Thromboembolic disease 48
Thrombophlebitis 48, 70, 74
Thrombosis 36

Thyroid function 78
Tooth disorder 69
Transdermal
    estradiol 51
    estrogen-progestin contraception 39
    patch 65
Trimegestone 9, 34
Triphasic pill 63

U
Urticaria 68
Uterine 23
    bleeding, abnormal 16, 35, 49, 52, 80
    fibroid 67
    volume 49

V
Vaginal
    contraception 37
    discharge 67
    dryness 67
    estrogen-progestin contraception 37
    gel 14, 18
    progesterone 55
    rings 35
    suppository 61, 82
Vaginitis 67
Verebrovascular disorders 74
Verruca 68
Viral infection 69
Vision, abnormal 68
Vomiting 68

W
Women's Health Initiative Memory Study 75

EU GSPR Authorised Reprsentative
Logos Europe, 9 rue Nicolas Poussin
1700, La Rochelle, France
Phone: +33 (0) 6 67 93 73 78
E-mail: contact@logoseurope.eu